I0091390

Scott Granville

Mapping the Geographical and Literary Boundaries of Los Angeles

Scott Granville

Mapping the Geographical and Literary Boundaries of Los Angeles

A Real and Imagined City

VDM Verlag Dr. Müller

Imprint

Bibliographic information by the German National Library: The German National Library lists this publication at the German National Bibliography; detailed bibliographic information is available on the Internet at http://dnb.d-nb.de.
   Any brand names and product names mentioned in this book are subject to trademark, brand or patent protection and are trademarks or registered trademarks of their respective holders. The use of brand names, product names, common names, trade names, product descriptions etc. even without a particular marking in this works is in no way to be construed to mean that such names may be regarded as unrestricted in respect of trademark and brand protection legislation and could thus be used by anyone.

Cover image: www.purestockx.com

Publisher:
VDM Verlag Dr. Müller Aktiengesellschaft & Co. KG, Dudweiler Landstr. 125 a, 66123 Saarbrücken, Germany,
Phone +49 681 9100-698, Fax +49 681 9100-988,
Email: info@vdm-verlag.de

Copyright © 2008 VDM Verlag Dr. Müller Aktiengesellschaft & Co. KG and licensors
All rights reserved. Saarbrücken 2008

Produced in USA and UK by:
Lightning Source Inc., La Vergne, Tennessee, USA
Lightning Source UK Ltd., Milton Keynes, UK
BookSurge LLC, 5341 Dorchester Road, Suite 16, North Charleston, SC 29418, USA

ISBN: 978-3-639-02553-8

ABSTRACT

In Los Angeles, the influence of Hollywood and the film industry, combined with a non-stop barrage of media images, has blurred the line between the real and imaged. The literature reveals a city exploding with cultural, racial and social differences, making Los Angeles a confusing and alienating place. The literature of Los Angeles reflects the changing face of the city. Los Angeles was always a city with a promising future, economic booms and optimism seemed to suggest that here was a place where the American Dream really could come true. Thousands travelled west in search of sunshine, oranges and a life that formerly, they could only dream of having. Yet, the literature of Los Angeles has highlighted the city's actual history together with a realization of undercurrents of violence, prejudice, depression and shattered dreams. The past, present and future is used to reveal a city that is in stark opposition to the Los Angeles, waves of immigrants came to find.

This thesis explores the idea of the dreamer coming west to Los Angeles within the literature and the variety of ways in the travellers' romantic notions of Los Angeles as a city of promise, is betrayed, leaving a desperate people in its wake. The literature shows that beneath the shiny surface of a city founded on sunshine and prosperity, corruption reached all levels of society and the 'mean streets' abound.

Later, influenced by an overwhelming feeling of powerlessness caused by Post-war nuclear depression, McCarthyism, loss of identity, and living in a city fragmented by racial tension and an ever growing gap between the very rich and the very poor, the literature of Los Angeles reflects not only the fears of that city, but of American society as a whole. The collision of technology, rapid progression and population explosion turned Los Angeles into a disconnected city, where the real and imagined merge in a cityscape that demonstrates a conflicting combination of historical replication, original design and movie-set inspiration. Nothing is ever what it appears to be in Los Angeles.

ACKNOWLEDGEMENTS

First and foremost, I would like to thank Dr. Jan Pilditch for her guidance and expertise. Without her help this journey would have been considerably more arduous. I would also like to thank Dr. Mark Houlahan for his words of encouragement and support that have always been greatly appreciated. Also, to the faculty and administration in the Arts and Social Sciences Department, I would like to thank you for the generous Scholarship Award and for all of your support during the course of the last year. Lastly to my family and friends, thank you for your patience and understanding. I cannot begin to express how much they mean to me.

TABLE OF CONTENTS

INTRODUCTION

When Cecilia Brady looked out of her window on the descent into Los Angeles in the beginning of *The Last Tycoon*, she saw "the California moon was out, huge and orange over the Pacific."[1] The moon illuminates the city below, and serves as a beacon to the dreamers and seekers that come to Los Angeles searching for success. The image is deceiving. The moon does not guide people towards paradise: Los Angeles is a violent, confusing, and final resting place.

The city of Los Angeles is a contradiction, as much an imagined city as a real one. For as many people who have believed in and chased the dream of success, there are as many who have been destroyed by the illusion of opportunity. Timothy Tangherlini comments:

> The object of fierce derision and intense glorification, Los Angeles can be viewed as a constant and ever-changing series of contradictory interpretations of space: for some, it is a city of dreams and for others, a city of despair; it is at once a city of extraordinary wealth and a city of crushing poverty, a city of culture and a city of plastic, a city of WASPs and a city of immigrants.[2]

This collision of extremes; economic, cultural and spatial create a city with over-lapping boundaries. It is a city built on an imagined past of harmony and prosperity, hurtling towards a future overflowing with people, each with a separate and conflicting dream.

Los Angeles has had a varied history. On July 6, 1769, Father Junipero Serra founded San Diego de Alcaca, the first of the Californian missions, following earlier established missions in New Mexico and Arizona. As George E. Tinker comments in *Missionary Conquest*, "Serra conducted his evangelist outreach to native people as part of the occupational force of a conquering army and as part of a strategy of conquest."[3] The historical fact of this missionary movement, protected by the Spanish Army, was chiefly responsible for the decimation of the local Native Indian population. This fact was completely ignored by the advertising propaganda of Southern California of the 1880's. In the advertising of the day, the missionaries and mission buildings were a focal point for a new campaign to bring people west. The buildings, run-down and decrepit, became an architectural symbol within the myth of Los Angeles in which the missionaries were revered as saviours, not repressors. David Fine comments:

> The forced enclosure of the mission Indians and their consequent decimation by cultural uprooting and European disease found no place [...] in this mythic reconstruction of history, which centered on the image of the humble, godly, kneeling countenance of the Spanish padre ministering to the needs of the ignorant, helpless aborigines...[4]

Growth, not historical accuracy was the order of the day and those promoting growth were known as boosters. The boosters of Southern California, who projected the idea of expanding the Los Angeles area at the turn of the nineteenth century, used opposing visions of progress and nostalgia, future possibilities and historic charm. The alluring combination of a frontier town, still echoing with the recent influence of the Spanish missions, alongside the commercial potential of a town recently connected to the major Pacific railway lines. It was an appealing vision.

With the mythical history established, the boosters undertook a major national advertising campaign, hawking consistently warm weather, healthy dry air and agricultural opportunity. These agricultural opportunities had not always existed. Mike Davis, in *Ecology of Fear* comments:

> It was not until the discovery of great artesian basins – millions of acre feet of subterranean water – during the 1870's, and the subsequent growth of the citrus industry, that it became possible for an Edenic vision of Southern California to bloom uncontested.[5]

For many people this was exactly the opportunity they were looking for. The vision of a landscape blanketed in citrus trees and smiled upon by constant sunshine offered an escape from their existing way of life. For the adventurer and the dreamer it was a golden chance to strike out for new territory and stake a claim.

Los Angeles was a place for new beginnings. It had commercial and health benefits to compliment the glorified past of a romantic and pastoral setting. These brightly commercialized ideals of financial opportunity and personal happiness, along with accessible trade and transportation routes brought a steady stream of settlers to the city of Los Angeles during the early part of the twentieth century. By the early 1920's, "a population that only thirty years earlier had numbered 100,000 now matched San Francisco's population of 575,000."[6] A growing emphasis on big business and a correlating boom in several industries fed the growth. Oil deposits were found in several locations close to, and within, the growing city limits. The aviation industry, itself once a dream scattered across several eastern states, found a home in Los Angeles due to the ideal climate and unlimited days of sunshine. As Edward Soja and Allen Scott comment:

> The Los Angeles aircraft industry also boomed in the interwar years, although it required the impetus of the Second World War before it assumed unquestionable national leadership, and it became the conduit through which the main high-technology industrial base of the regional metropolis would be established in later years.[7]

Los Angeles was alive with creative, analytical and entrepreneurial spirit. Nothing appeared insurmountable, there was vibrancy to the city and people became caught up in the idea that Los Angeles did indeed offer everyone the opportunity to succeed.

Feeding the dream was another industry that had found a home in Los Angeles; the film industry. While the original dream for Hollywood was not in film but an attempt at a Christian utopia by wealthy Kansas real estate developers during the 1890's, after its failure, the natural beauty of the area made it suitably attractive for the burgeoning film industry to take over. Colin Shindler comments:

> What lingered after this first dream died was the sensation that Hollywood still offered a geophysical paradise. In the very first days of the pioneers the air was attractively temperate, fragrant with orange blossom, jasmine and eucalyptus. Even in 1929 it was still possible to be astonished by the bunches of red berries on the ubiquitous pepper trees which lined the major roads.[8]

The original pure-thinking ideals did not last long in Hollywood, but the temperate conditions, abundance of sunshine and proximity to the Pacific Ocean made Los Angeles an attractive location for the film industry.

With the addition of sound to the movies of the late 1920's, a new generation began to arrive in Los Angeles. Established writers, promised high salaries and steady work, came to Hollywood to develop and write film scripts. While the money was often generous and the demand for scriptwriters high, many of the writers were disillusioned by their experience. Authors such as James M. Cain, Nathanael West and F. Scott Fitzgerald were among the first to write about events that not only questioned the El Dorado myth of Los Angeles as a golden city but also completely rejected it. Fitzgerald who had come to Hollywood on three separate occasions in the hope of resurrecting his career wrote:

> The picture I have just finished is in production and though it bears my name, my producer could not resist the fascination of a pencil and managed to obviate most signs of my personality. Nonetheless I am now considered a success in Hollywood because something

which I did not write is going under my name, and something I did write was quietly buried...[9]

The contradictory and fickle nature of the film industry, together with the geographical and historical pretences that established Los Angeles as a destination, provided the basis for many of the literary themes that grew out of the city. David Fine comments:

Los Angeles could not serve as setting for the regenerative possibilities of America. The dream, if it had once had potency was behind them. Writing against the myth of El Dorado, they transformed it into its antithesis: that of the dream running out along the California shore[10]

The landscape, seen through the eyes of its writers, transformed Los Angeles from a bright, sunny place of opportunity into a dark, at times farcical and at times hysterical, nightmare. The scenery and architecture that gave Los Angeles its identity; the ocean, the road and the buildings became symbols for the impending chaos of a city on the brink of disaster. In the writing of these authors, beginning in the 1930's, the darker side of Los Angeles was exposed, in ways, far removed from the rejuvenating, exotic images that had been implanted by the boosters.

When America entered the Second World War in 1941, Los Angeles experienced a massive influx of migrants to service the war effort through the growing aeronautics and munitions industries. A city resigned to, yet thriving on, an unprecedented urban sprawl, brought about through a ban on buildings above fifty metres in height and a maze of criss-crossing freeways, continued to promote the virtues of an imagined paradise, and to ignore the growing disparity in wealth and opportunity. Kazys Varnelis comments:

Los Angeles swiftly evolved its own ideology of urban growth, combining the promises of progressive urbanism with dollar-driven planning. Together with city government officials, whose elections they bankrolled, developers downplayed the center city, promoting instead the idealized image of Los Angeles as a place in which prairie and city life collapsed into each other and the ideals of the garden city could be realized.[11]

The reality of this idealised portrayal became not the unrestricted, co-existing paradise it promised, but a fragmented, disjointed city run by corrupt city officials. Fuelled by these developments, the paranoid activities of a nuclear wary Cold War Government, and the hypocritical presentation of the Hollywood film industry, literature in Los Angeles turned against the media-driven image of unity and fulfilment. The line between fiction and reality blurred as contradictory notions of architecture tried to reconstruct historical buildings while embracing modern techniques. Los Angeles streets became Hollywood sets, confusing the imagined view with real life. Using the past, present and the future as reference points, the writing exposed the feelings of desperation and isolation experienced by many people as they searched for meaning in a city built on a history of false promises and dreams. This thesis will examine the literature of Los Angeles in the light of the city's development. In Chapter One, the idea of the dreamer coming west to Los Angeles is explored. The romantic notion of Los Angeles as a city of opportunity is dismantled, revealing the desperation of people whose dreams can never be realised or satisfied. Chapter Two discusses the physical expansion of Los Angeles following the Second World War and how the disconnected, sprawling nature of the city reflects on the literature. The effects of the Cold War and nuclear paranoia also begin to reveal a new sense of isolation and confusion within the individual, and the writing begins to challenge society's values. In Chapter Three, the popular image of Los Angeles is challenged, exploring the vastness of a diverse cultural, racial and economic city. The sun-kissed, sandy beaches of Malibu are replaced by a city filled with violence, misinformation and uncertainty.

---

[1] F. Scott Fitzgerald, *The Last Tycoon* (Harmondsworth, UK: Penguin Books Ltd., 1977), p. 25.
[2] Timothy T. Tangherlini, 'Los Angeles Intersections (Folklore and the City)', *Western Folklore*, 58.2 (1999), p.99, in JSTOR <http://links.jstor.org/sici?sici=0043-373X%28199924%2958%3A2%3C99%3ALAI%28AT%3E2.0.CO%3B2-1> [accessed 11 December 2006].

[3] George E. Tinker, *Missionary Conquest* (Minneapolis, MN: Orca Press, 1993), p. 44.
[4] David Fine, *Imagining Los Angeles* (Albuquerque, NM: University of New Mexico Press, 2000), p. 30.
[5] Mike Davis, *Ecology of Fear* (New York, NY: Vintage Books), p. 11.
[6] David Fine, *Imagining Los Angeles*, p. 4.
[7] Allen J. Scott and Edward W. Soja, *The City – Los Angeles and Urban Theory at the End of the 20th Century* (Los Angeles, CA: University of California Press, 1996), p. 7.
[8] Colin Shindler, *Hollywood in Crisis* (New York, NY: Routledge Press, 1996), p. 2.
[9] *Correspondence of F. Scott Fitzgerald*, ed. by Matthew J. Bruccoli and Margaret M. Duggan (New York, NY: Random House Inc, 1980), p. 176.
[10] *Los Angeles in Fiction: A Collection of Essays Revised Edition,* ed. by David Fine (Albuquerque, NM: University of New Mexico Press, 1995), p. 7.
[11] Kazys Varnelis, 'Los Angeles – Cluster City', in *Future City*, ed. by Stephen Read, Jurgen Rosemann and Job van Eldijk (Oxon, UK: Spon Press, 2005), pp. 174-93.

8

CHAPTER ONE

Versions of the American Dream are as various as the people who come here in search of fulfilment. Yet every dream, it seems, has a similar sequel, one that spells disillusion.[1]

The I-10 highway heading west crosses the San Bernardino county-line, by-passes the towns of Fontana, Pomona and Redlands and cuts through the heart of Chinatown and eventually finds Santa Monica. At this point the highway goes no further west. In front lies the glittering expanse of the Pacific Ocean, behind a city built by over a hundred years of dreamers, seekers and travellers who all came west at one stage in search of opportunity. This road has always appealed to the adventurer, the traveller, the dreamer. There is a sense of freedom associated with travel that arouses the possibility of opportunity. The dream of opportunity has at many times led the traveller to Los Angeles as a destination. It is not an easy city to reach, the highways cross desert or mountain ranges, the coastal roads traverse a collection of peaks overlooking the Pacific. On arrival in the city, the newcomer senses an undeniable truth, that there is nowhere further west to travel, that the final frontier has been reached.

The Southern Californian desert, together with the surrounding mountains encases the city of Los Angeles, but it is a city that continues to grow. Mike Davis in *City of Quartz* comments that the desert is, "being prepared like a virgin bride for its eventual union with the Metropolis: hundreds of square miles of vacant space engridded to accept the future millions"[2] As the known residency of the municipality runs over the 15 million inhabitant mark and inflation "reduces access to new housing to less than 15% of the population,"[3] what was once uninhabitable desert land now provides the latest opportunity for the new wave of investors, families and dreamers. The surrounding desert space that once kept people away from Los Angeles now, through the benefits of modern engineering, allows the creation of new sub-divisions. Nevertheless, the Pacific Ocean remains the immovable mass that defines the physical end to the road heading west. If the recreational and health purposes of the ocean appear to be appealing and beneficial, on the outside, the presence of the ocean has often been used by the writers of Los Angeles as a symbol for the death of the American Dream and the end of the road.

James M. Cain's *The Postman Always Rings Twice* marks the beginning of the Los Angeles novel of this type and the novel works to deflate the image of Southern California as the land of opportunity. Using the highway and the roadside Twin Oaks Tavern as central metaphors, the contrasting Californian dreams of mobility and domesticity are dismantled. The use of a violent, adulterous affair leading to murder as the principal storyline only reinforces the emerging belief that Los Angeles as a city was slipping into a dangerous future. The road and the car cease to symbolise mobility, new beginnings and freedom. Rather, these symbols become the focal point for death, destruction and the loss of hope. Even as Frank Chambers, a serial drifter and the narrative voice of the novel, enters the roadside Twin Oaks Tavern for the first time, the highway remains his escape route. "I blew in there in a hurry and began looking down the road."[4]

Frank has been living on the road for so long he is always keeping one hand on the door, ready to leave at a moments notice. But when Frank meets Cora for the first time and realizes, "I wanted that woman so bad I couldn't even keep anything in my stomach,"[5] he

finds his beliefs and way of life compromised. Cora, wife of the proprietor Nick is herself

conflicted; with the dream of wanting domestic bliss and her utter contempt for her

husband. David Madden comments on Frank and Cora:

> Their extreme insecurity accounts in part for their lack of conscience. Frank, the drifter lost
> in a land of promise, no longer seeks the dream; and he is compelled to act the nightmare
> when he can no longer elude it. In the background, Frank and Cora pursue separate dreams
> which mock the shared realization of the immediate wish.[6]

Their mutual attraction is immediate; Cora views Frank as the perfect ally in murdering her

husband Nick, while Frank's thoughts towards Cora are physical and lustful. He needs to

'have her.' Yet the shared physical lust is not powerful enough to overcome their

conflicting dreams. The instantaneous gratification they feel for each other cannot mask

the reality of what the future holds, an unfulfilled and tragic ending. After a first failed

attempt at murdering Nick, Frank turns to the road as an escape. Planning to leave before

Nick's return from hospital, Frank says it will be, "just you and me and the road, Cora,"

romanticising the idea by adding that they would be, "just a couple of gypsies but we'll be

together."[7] Cora realizes living on the road is not her dream and will not continue with the

plan. Her overpowering drive is her need for security but she also desires the glamour and

excitement that brought her to Los Angeles. What Nick lacks in terms of providing a

glamorous and exciting life, he more than makes up for in dependability and financial

security. Frank too is a creature of habit, he is a serial drifter. He returns to the road to

avoid responsibility for his actions.

The constant struggle faced by Cora and Frank to realise their dreams is reflected

throughout the writing of the Depression era. Americans from across the country were

affected by the economic recession and Los Angeles appealed as a city with a future.

"They took a look at the tormented land and overburdened cities of their own regions," and directed, "their families across the plains and deserts to the golden valleys of the West

Coast."[8] On the surface Los Angeles appeared to be a place that allowed dreams to come

true. Once in the city however, many of the dreamers found their hopes washed up against

the shores of the Pacific Ocean. Lee Horsley comments:

> Cain presents his characters as victims traumatised by national economic disaster but
> nevertheless driven by myths of limitless opportunity, success and unhampered self-
> determination. They follow the ignis fatuus of the American dream, and when they have
> (opportunistically) attained their wishes they find that all they have really secured is defeat
> and entrapment.[9]

The freedom Frank, in *The Postman Always Rings Twice*, offers through life on the road

cannot compete with the powerful draw of potential Cora sees in the Twin Oaks Tavern.

Cora's earlier dream of Hollywood happiness has been tempered by her new goal, a

successful business and home that allow her to escape from financial desperation.

While Frank and Cora are separated for a short period of time, there is inevitability, a

certainty they will be reunited. Like the writing of earlier authors such as Frank Norris'

McTeague and Theodore Dreiser's *Sister Carrie*, there is an underlying Naturalism that

suggests man's actions are determined by sociological conditions and characters lives

governed by forces of heredity and environment. June Howard comments:

> The desiring, endeavouring self confronts an indifferent impersonal field of force that seems
> to have nothing to do with, to utterly negate – to be the contrary of – individual hopes and
> efforts.[10]

The force of nature is often too powerful to fight against and the individual can only

submit. The biological and psychological factors that make up Nick and Cora's characters

push them to an inevitable end. Frank's acceptance of Nick's invitation to return to the

Twin Oaks Tavern makes his reunion with Cora unavoidable and reiterates the effect of

everything happening twice in the novel; illicit affairs, murder attempts, escapes, and finally, deaths. Cora, dragged into her old nightmare of a domestic prison, is determined to finish what she had attempted once before with Frank, the plan to murder her husband. Frank agrees saying, "Cora, it's in the cards. We've tried every other way out."[11]

The road now transforms from a symbol of limitless opportunity into a dark accessory to murder. On the winding roads surrounding a Malibu beach, Frank and Cora stage the accident that will kill Nick. After listening to his dying echoes reverberate through Glendale Canyon and running the car off the edge of a ravine, Cora and Nick are possessed by an animal-like desperation of passion. "Hell could have opened for me then, and it wouldn't have made any difference. I had to have her, if I hung for it. I had her."[12] They make love beside the destroyed vehicle containing the recently murdered body of Nick, their clothes tattered and bodies bruised. The turbulent, fast-paced nature of the action drags Frank and Cora along, they are merely passengers aboard something much larger than anything they can control, momentarily losing the will to fight against the forces of nature as they succumb to their desires. The road, which earlier offered so much in the dream for mobility and freedom has collapsed into a metaphor for futility. Frank and Cora escape punishment for murdering Nick only for Cora to die in a car accident. Frank is sentenced to death for killing Cora and lives out his final hours with the understanding he has not escaped anything but has been returned by the road to the site of his original wrongdoings.

Cain's depiction of Los Angeles in *The Postman Always Rings Twice* exposes the city's darker side. The emptiness shown in the ending of the novel, with the death of Cora in a car accident and Frank awaiting the death penalty for her murder, marks the beginning of hard-boiled fiction writing in Los Angeles during the 1930's. The peeling away of society's superficial layers reveal a bleaker, more cynical outlook on the city and is associated with the establishment of the detective fiction writing beginning with the popular *Black Mask* detective serials, the novels of Dashiell Hammett, and later Raymond

Chandler. Like James M. Cain's Frank Chambers, in Hammett's *The Maltese Falcon* and

Chandler's *The Big Sleep*, the protagonists protect themselves by emotionally insulating

against society, creating the tough-guy image synonymous with hard-boiled writing. Paul

Skenazy comments:

> Although invariably delighting in the pleasures of power and manipulation, the detective is
> finally able to reject temptation. He declines the achievement-orientated, financially
> defined power structure of American society, opting instead for his own independent point
> of view.[13]

The hard-boiled protagonists of Hammett and Chandler are a new breed. Determined to

maintain their own individualistic code of honour and justice they are city savvy detectives

aware of the corruption on all levels of society. They make their living mixing with the

criminal element on the new 'mean streets' of Los Angeles, which, at the hands of

Hollywood, has become all of California's and America's cities.

Although *The Maltese Falcon* is set further north in the city of San Francisco, the

hard-boiled nature of Hammett's writing resonates with the Los Angeles detective image of

one man fighting against the corruption of the world, establishing and reinforcing the

principles seen in the writing of Raymond Chandler and other Los Angeles detective

writers.. Porter comments:

> The underlying dialectic in their novels reveals the ongoing tension between the individual
> – the hard-boiled dick – and the environment created by efforts to implement the dream of
> the Golden West. The dream itself is seldom mentioned in these novels, and there is little
> reason to mention it, because the world pictured is a measure of the stark hiatus between the
> dream and the reality.[14]

The life picture described by Hammett and the hard-boiled writers who would succeed him

is one of ambiguity and tension, of the individual's struggle to live in a world of corruption,

deceit and failed dreams. It would become a central idea in the art form of film noir that flourished during and after the Second World War, particularly with John Huston's adaptation of Hammett's *The Maltese Falcon* (1941). Gary J. Hausladen and Paul F. Starrs comment:

> The key characteristic of film noir is ambiguity, especially in the triangular relationship between a male protagonist, a femme fatale, and a crime: ambiguous protagonists, ambiguous femmes, and ambiguous victims. The tie, therefore, between film noir and Los Angeles is ambiguity and the unsettled city; impermanence is much of what classic noir deplores.[15]

It also meant, however, that Los Angeles became to the film-going public, every city.

The California lived in by both Sam Spade in Hammett's *The Maltese Falcon* and Phillip Marlowe in Chandler's *The Big Sleep* is considered the new frontier, a symbol for renewal and prosperity but undermined by corruption. This tension creates a darker side of society, exposing the dreamers who had not fulfilled the expectations they had set for themselves, as they looked to unlawful enterprises as a new route for success. Joseph Porter comments:

> The hopes projected for California were, for some, realised. For others, California was measured by the expectations it failed to fulfil and was judged accordingly. California and its wider reflection, the American West, is thus to some degree an expression of the tension between the cherished hopes and the disappointment of reality.[16]

Los Angeles had become the city at the end of the American West and failed dreams could no longer be left behind or ignored. Scratching the surface, beneath the golden exterior of sunshine and oranges on the California coast was the desperately realistic struggle for every-day survival.

The character of Sam Spade is a re-invented breed of American hero, an individual willing to manipulate society's conventions and norms to maintain his place in the universe. However, there is an ambiguity about Spade that never really separates him from the criminal element he lives among, leaving the interpretation open as to whether or not he works to ensure justice prevails or simply for monetary reward. Lee Horsley comments about Hammett:

> He introduced characters who much more nearly conform to the description of the private eye as 'half gangster' – a man whose innocence has become so tarnished as to be no longer visible, and who is a close relation of the crook-as-investigator protagonists who emerge…in the early thirties.[17]

Even Spade's physical appearance suggests something darker, "He looked rather pleasantly like a blond Satan."[18] The ongoing conflict between protagonist and accepted behaviour, coupled with the third person narrative which allows the withholding of important facts, creates the tension that moves the story forward. He uses his knowledge of the law to ensure he stays one step ahead of his competition. Spade acknowledges the corrupt world he lives in, realises the importance of understanding people's intentions, and uses this intuition to see through their deception.

The value of human life and the uncertainty of an individual's place in the world written about by Dashiell Hammett are encapsulated in the story told by Sam to Brigid O'Shaunessy involving the Flitcraft case he once worked on. Flitcraft is a man who escapes a beam falling on his head by mere seconds and realises the fragile nature of man's time on earth. He disappears, leaving a family and an established life, reappearing years later under a new name with a new life and family. Cynthia Hamilton comments:

> The Flitcraft anecdote in *The Maltese Falcon* perfectly encapsulates the bitter irony of what Hammett sees as the human condition: Flitcraft assumes not only a universe structured on human notions of rationality and justice, but also a correspondence between his perception of the structural rules and the actuality.[19]

Hammett and his character Sam Spade, like Flitcraft, realise the precarious nature of any individual's place in the world. The recognition of life's impermanence leads people to make self-centred decisions, often in conflict with their immediate community. There are no fairy-tale endings and life continues to be harsh and unpredictable. Los Angeles would only serve as further proof to this belief as the dreamers who came to the city soon realised there was nowhere left to escape. More problems are certain to arise and the only person left to trust on the 'mean streets' becomes yourself.

The influence of Dashiell Hammett's writing on Raymond Chandler and his own hard-boiled detective, Philip Marlowe, is indelible. Frank McShane comments:

> Chandler's admiration for Hammett was based on two related features of his work – his subject matter and his language. "Hammett," he wrote, "took murder out of the Venetian vase and dropped it in the alley." Unlike English detective stories in which murder was an affair of "the upper classes, the weekend house party and the vicar's rose garden," Hammett "gave it back to the people who commit it for reasons, not just to provide a corpse. He put these people down on paper as they were, and he made them talk and think in the language they customarily used for such purposes.[20]

The realistic, gritty endeavours of Hammett's characters now applied to the streets of Los Angeles and Chandler's own California experience with the oil business and living in the Depression era. Chandler could also draw on the exposure of the political corruption running rampant in the Los Angeles city government and the much publicised nomination of author Upton Sinclair to run for the Democratic Party in the gubernatorial race of 1934.[21] Realising the potency of Hammett's writing style, Chandler's brand of hard-boiled detective story is a commentary on the corruption that flourished and fed off people trying to avoid the reality and harshness of life while still trying to fulfil their dream. McShane writes:

> The law was something to be manipulated for profit and power. The streets were dark with something more than night. The mystery story grew hard and cynical about motive and characters, but it was not cynical about the effects it tried to produce nor about its technique of producing them.[22]

If Hammett's influence in subject matter is evident, Chandler chooses to move away from the third person narrative and use his character of Philip Marlowe in first-person narration. Starting with *The Big Sleep* in 1939, the first-person narrative technique allows Marlowe to develop a distinct, identifiable personality. Sam Spade is tough, clever and street-smart but it is with Marlowe that the reader begins to connect with the observations, the emotions and the dislikes of the hard-boiled detective on a more intimate level.

In *The Big Sleep*, no social class is exempt from the corruption and deceit on the Los Angeles streets. More than any of his other works, *The Big Sleep* is an anti-rich novel, where the carelessness and irresponsibility of the wealthy is exposed. The opening of the novel is framed around the huge Sternwood mansion, where Marlowe has had his services requested by the ancient, ailing General Sternwood. The house overlooks the oilfields that have made the family fabulously wealthy.

> The Sternwood's, having moved up the hill could no longer smell the stale sump water or the oil, but they could still look out of their front windows and see what had made them rich. If they wanted to. I don't suppose they would want to.[23]

Their money has been made and the Sternwood's can enjoy the benefits of financial security. While the house is physically above the lower reaches of the city below, the family cannot remain detached from people who live in it. Chandler's *The Big Sleep* reveals that no amount of money can cover up and protect the unlawful activities and social problems of the Sternwood family.

Both daughters of General Sternwood are living lives filled with lies, secrets and deceit. The elder daughter, Vivian, is married to an Irish bootlegger who has mysteriously disappeared and the younger daughter Carmen is heavily involved in drugs, running up debts for her father and making herself the unsuspecting victim of several blackmail scandals. Marlowe's introduction to Carmen early in the novel provides the catalyst for shaping her character. "Then she lowered her lashes until they almost cuddled her cheeks and slowly raised them again, like a theatre curtain. I was to get to know that trick. That was supposed to make me roll over on my back with all four paws in the air."[24] Her girlish coltishness and the theatrical battering of eyelashes is a cosmetic covering for her insecurity and immaturity. Like Sam Spade does to Brigid, Marlowe is able to resist Carmen's flirtatious actions, alert to the false intentions attached to them. He becomes the ultimate impartial examiner, not judging by appearance or status but through his methodical collection of information. Empowered by the romantic ideal of sole protector of justice, Marlowe remains separate, judging from an independent standpoint. Meagan Abbot comments:

> Marlowe is positioned explicitly against any system – juridical of baldly capitalistic. Ostensibly a deductor as a detective, he is however, not a slave to the system of knowledge and its seeming valorisation of logic and reason. Instead, Marlowe lives by what he considers his own "code," his moral ideology and he gets no "kick" from violating it through the familiar generic vehicle of transgression: a heterosexual liaison with the femme fatale.[25]

Marlowe is able to overlook the almost immediate advances of Carmen Sternwood in order to continue his paid task for "twenty-five a day and expenses."[26] He is opposed to the carelessness Carmen shows in her on-going advances to him. When later he finds her waiting naked in his apartment, Marlowe is restrained, telling her to get dressed. When he gives Carmen a final rejection, she curses at him. Only then does Marlowe lose some of his composure. "I couldn't stand her in that room any longer. What she called me only reminded me of that."[27] Marlowe is protective of his environment, his apartment a sanctuary away from the harshness of the Los Angeles streets and he will not allow Carmen to pollute this safe haven.

Unlike the criminal group including Gutman and Brigid O'Shaunessy in *The Maltese Falcon*, the Sternwood's social status should elevate them above any kind of illegal activity. Yet the superficial nature of their standing in Los Angeles' upper-class, identified with Vivian Sternwood, is exposed as a fraud. There are gambling problems and close associations with members of the criminal underworld. *The Big Sleep* is more than a detective novel: it is a comedy of human futility, where the wealthy are careless with the power they have and use it to mistreat not only themselves but also the people around them. Unravelling the American Dream, *The Big Sleep* shows criminal activity infiltrating even the most privileged families, mocking the publicised version that crime was restricted to low socio-economic communities in the city.

Los Angeles, during the Depression, exemplified the state of the country in general. Middle-class savings had been poured into oil speculations and real-estate fluctuations affected people with as much insecurity as anywhere else. Framed around a marathon dance competition on the pier in Santa Monica and the sentencing to death of the protagonist Robert Syverton, Horace McCoy's, *They Shoot Horses Don't They?* is a further examination of futility and the betrayal of promise that characterises the Depression era in Los Angeles. McCoy's use of the dance competition in a hall over the Pacific Ocean reiterates the image of the highway's end as the end of the dream and the circular track used by the competitors becomes an absurd expression of the continuity in the search of promises left unfound. McCoy's Robert Syverton and his partner in the competition, Gloria Beatty, are both Los Angeles outsiders. They have been seduced, at separate and unrelated times, by the lure and potentially instant success of Hollywood. Robert's ambition and dream lies in directing movies, and unlike Gloria, he has not lost the romantic ideals associated with the opportunity of success. David Fine comments:

> The contrast between Robert and Gloria is that between traditional native optimism and
> absurd existential awareness. Robert brings to the West Coast an innocent hopefulness, a
> literal belief in the Gospel of Hollywood; and she the depilating awareness that the world
> doesn't square with hopes, logic, or conventional morality.[28]

Robert, although struggling to make an impact still has the desire and belief that his dreams can be manufactured into reality. Gloria has come to Hollywood with a different dream, to escape her past. After running away from an uncle who made sexual advances on her, stealing from a store and being arrested, before living with a Syrian, who, "wanted to make me between customers, on the kitchen table,"[29] Gloria takes poison, wanting to die. When she doesn't take enough and ends up in hospital, Gloria reads about Hollywood from movie magazines and decides to try to become an actress. This desire stems more from having nothing else to turn to than from any burning passion to act. Her nihilistic view on life and obsession with death are in direct opposition to Robert's outlook before the dance. Gloria tells Robert, "if I had the guts: I'd walk out of a window or throw myself in front of a streetcar or something."[30] Nothing Robert can say will alter Gloria's view of her position in the world.

The dark side of society is always close to the surface in *They Shoot Horses Don't They*? The competition becomes a metaphor for the dance of life, where the good is mixed with the ordinary and the downright bad. David Fine comments:

> The marathon itself is pure theatre – an elaborate, staged spectacle cynically manipulated by
> its criminal promoters to draw crowds of thrill seekers – and a parody of the Hollywood
> dream factory.[31]

A contestant with whom Robert has become friends, Mario, is arrested after it is discovered he is an escaped murderer. "He was one of the nicest boys I'd ever met. But that was then I couldn't believe it. Now I know you can be nice and be a murderer too."[32] There seems to be no escaping the circular nature of life. Able to reflect back as his own sentence is passed down, Robert understands an individual action does not define a person as a whole. It is simply a single incident, one event in a random set of events that make up the universe. The irony comes in Mario's identifiers, two detectives taking time off to watch the dance competition and recognise his face from a popular crime-fighting magazine. The detectives are only a small part of the spectators, a collection of celebrities, criminals and everyday people, joining together to encourage the contestants in their desperate struggle for victory in a competition that goes nowhere.

The dance hall set out over the Pacific symbolises the end of the highway, the end of the dream. Trying to escape from the harsh reality of a desperate life, the competitors fail to realise that they are travelling on a circular path. Fine comments:

> That McCoy's novel is set not only against the ocean but over it as well, on the pier in Santa Monica, intensifies this sense of being at the very end of the road, the place from which there is no return. The back-and-forth movement of the dancers across the floor – movement without progress – is accompanied by the monotonous, relentless ebb and flow of the ocean beneath their feet.[33]

As the competition continues, the physical exhaustion adds to the futility of the dancing. The dancers pound their way around the dance track, going nowhere, they are entertaining but not achieving anything tangible. During one of the short breaks allowed to competitors, Robert is convinced to go under the announcement platform with a female contestant. As his heart starts pounding, Robert has a startling revelation. "There is no new experience in life. Something may happen to you that you think has never happened before, that you think is brand new, but you are mistaken."[34] Gloria's nihilistic outlook begins to make some sense to Robert. He sees there are no new experiences in life. Robert's path in seeing the deceitful nature of life is gradual.

Early in the dance Robert is still enlightened by the ocean and its natural beauty. Looking out of the door he witnesses the sunset. "Out there where the sun was sinking the ocean was very calm, not looking like an ocean at all. It was lovely, lovely, lovely, lovely,

lovely, lovely."[35] Robert refers to the people fishing and not paying attention to nature's

beauty as fools, finding them ignorant of Los Angeles' redeeming natural qualities. Yet by

the end of the dance, the hopelessness and absurdity of the competition and its reflection on

life as a whole has altered his thinking. Lee Richmond comments:

> It is through the image of the eternal, meaningless dance – with its reminder of the futile
>
> rhythm of existence – that Gloria and Robert are ultimately stripped of their particularity
>
> and become symbolic projections of the absurdist man and woman. Gloria's 'world pain' is
>
> seized by Robert in the throes of his dance movements: his physical exhaustion is a tangible
>
> correlative for his spiritual weariness.[36]

When all elements of life are put together, living and dying becomes a lottery, an arbitrary

act played out by a higher force than anything in this world. Robert tells Gloria, "Don't

think I'm crazy about this ocean," saying, "It'll be all right with me if I never see it again.

I've had enough of the ocean to last me the rest of my life."[37] He has lost his spiritual

connection with nature. What was once alluring and captivating, beautiful and precious, is

now stale. The dance competition, like life, has been drained of its romantic ideals.

Gloria, death obsessed and finally abandoning all hope of happiness, gives Robert

her pistol. Together on the pier, out over the ocean, the road is at its end. "Take it and

pinch-hit for God. Shoot me. It's the only way to get me out of my misery." Robert, tired

with trying to understand life, now agrees with her. "She's right. It's the only way to get

her out of her misery."[38] Detached from himself, Robert has come a long way towards

Gloria's own nihilistic vision. He shoots and kills her with a clear conscience. Like

Camus' Mersault in *The Stranger*, Robert accepts his sentence of death in a rational

manner, knowing that within society's rules he has committed a crime. At the same time he

justifies his actions, believing that his actions have released Gloria from the continuous cycle of a miserable and depressing life. Standing on the edge of the west, with Los

Angeles behind him, Robert's dream ends with Gloria's, in death.

The Los Angeles landscape, as seen through the eyes of outside writers brought into

Hollywood on the wave of the growing motion picture industry in the 1930's, transformed

the city from a land of opportunity into a living nightmare. The very scenery and

architecture that gave Los Angeles an identity, the sunshine, the highway and the ocean, all

became metaphors for the impending disaster of a city on the brink of social chaos.

Hollywood, home to producers, directors and film stars, seemed far removed from the

Depression and an escape for the dreamers and seekers who became central to the Los

Angeles novel. The film industry continued to expand, feeding off the principle ideas of art

and commercialism, both essential to its existence but conflicted in its ideals. Writing

about Hollywood, Raymond Chandler said:

> The story that is Hollywood will someday be written and it will not primarily be about
> people at all, but about a process, a very living and terrible and lovely process, the making
> of a single picture, almost any hard-fought and ambitious picture, but preferably a heart-
> breaker to almost everyone concerned. In that process will be all the agony and heroism of
> human affairs, and it will all be in focus, because the process will be the story. Everything
> that matters in Hollywood goes into this process. The rest is waste. Above all the vice is
> waste, and the vicious people, of whom there are many and always will be, because
> Hollywood is starved for talent, for a single facet of a single talent, and will pay the price in
> disgust, because it has to. [39]

Hollywood and Los Angeles are never far from the movies. The people involved in making

movies, both important and peripheral become part of their creation. There is the struggle

to find balance between artistic reward and financial gain, to avoid being swept up in the

superficial values of those who succeed and the desperation of the many who do not

succeed. From the vast back lots of the film studios to the street corners that double as part-time shooting locations, the physical impression of Hollywood blurs the fiction of what is being filmed into the reality of life until it becomes difficult to differentiate between the two.

Works like F. Scott Fitzgerald's *The Last Tycoon* and Nathanael West's *Day of the Locust* are commentaries on Hollywood and examine the development of the individual as they are influenced by the building up and dismantling of the California Dream. Christopher Ames comments:

> Hollywood fiction recapitulates the classic themes of American fiction in exaggerated form. The dream of westward migration to a richer land of opportunity and freedom, the dream of discovering a new Eden...the association of the westward rush with an imagined victory over mortality...[40]

Hollywood is essentially a 'dream factory', an industry designed to capture the imagination and promote the possibility of fiction becoming reality. Hollywood, having such an intimate relationship with the city of Los Angeles, worked to convince people through its images, to travel west in search of fulfilment. Such travellers soon discovered the harsh realities of the film industry, a hypocritical and contradictory world and succumbed to the truth of their own mortality and shortcomings.

Fitzgerald was looking for renewal when he wrote *The Last Tycoon*. His notoriously lavish partying lifestyle and the accumulating costs for his wife Zelda's health care left him in a financial crisis, the lure of large pay-cheques too enticing to reject for artistic independence. His two previous visits to Los Angeles had not been overly successful and his difficulties of adjusting to screen-writing are reflected in the number of writing credits, one, that he had achieved to date. With his own health failing, Fitzgerald looked to Hollywood as a last opportunity to have his own piece of the American Dream. K. G. Cross comments:

> That Fitzgerald – for whom the American Dream had a particular significance – should
> have come, at the end of his life, to live and write about the place where the dream is
> packaged and purveyed, has seemed to many commentators peculiarly appropriate, yet it is
> its verisimilitude to the Hollywood behind the greasepaint – the eczema of a actress's
> shoulders, the secretary thrust, naked, into a cupboard – that distinguishes his last book. [41]

In *The Last Tycoon*, Fitzgerald goes behind the 'grease-paint' and artificial movie sets of

the film world, identifying with the frailty of the human condition, and realizing through his

characters that power, success and influence cannot protect an individual from the

overpowering elements of a broken-heart or a shattered dream.

Monroe Stahr is a Hollywood outsider who at a young age works his way into an

executive position at a major studio through diligence, intuition for what works in films,

and being, "a money man among money men. Then he had been able to figure costs in his

head with a speed and accuracy that dazzled them…"[42] Based on Irving Thalberg, a Metro-

Goldwyn-Mayer executive who died at an early age, Stahr's character represents the

struggle to maintain integrity in the movie-making business at a time when the most

important line in the industry is the money line. Stahr is committed to making films with

substance, not always concerned with the total gross figure. In a conference for an

upcoming film, he admits the movie will lose money but says, "It's a quality picture," and

that, "we have a certain duty to the public."[43] Later, when he meets a man on the beach and

becomes involved in a conversation with him, the man tells him he never goes to the

movies. When asked why he replies, "There's no profit. I never let my children go."[44] The

profit he refers to is the intellectual meaning in a film, the learning experience, which the

man sees as non-existent. When Stahr realises this he changes the upcoming schedule of

shooting. "A picture, many pictures must be made to show him he was wrong."[45] He needs

to justify his involvement in the film-making process, to make it all seem worthwhile. Stahr is a moral bastion among the growing greed, shallowness and corruption of Hollywood and

Los Angeles. John Aldridge comments:

> Stahr, the last frontiersman, the embodiment of Fitzgerald's search for values beyond all
> frontiers, has come to rest at last in Hollywood, where the frontier has become a thing of
> cardboard and tinsel and the American Dream a corporation dedicated to the purveyance of
> dreams.[46]

While the hard-boiled detectives fought to keep the Los Angeles streets clean from crime

and corruption with their own brand of justice, Fitzgerald's Stahr fights his battles in the

Hollywood boardrooms, the executive offices and on the lots of a major movie studio.

The images from the city that Fitzgerald describes highlight the close interaction

between fiction and reality, dismantling the idea of truth winning out over disillusionment.

"Under the moon the back lot was thirty acres of fairyland – not because the locations

really looked like African jungles and French chateaux and schooners at anchor and

Broadway at night, but because they looked like the torn picture books of childhood, like

fragments of stories dancing in an open fire."[47] There is an undercurrent of the absurd as

people choose to compare movie sets with childhood memories, remembering the

innocence of their past lives rather than dealing with the reality of the present. The arrival

of Kathleen Moore, Monroe's love interest in *The Last Tycoon*, further captures the

physical presence of a city immersed in the production of movies. With Los Angeles

rocked by an earthquake, the studio back lots become flooded creating a surreal picture.

"On top of a huge head of the Goddess Siva, two women were floating down the current of

an impromptu river. The idol had come unloosed from a set of Burma, and it meandered

earnestly on its way, stopping sometimes to waddle and bump in the shallows with the

other debris of the tide."[48] Monroe becomes transfixed by Kathleen. "Smiling faintly at him

from not four feet away was the face of his dead wife, identical even to the expression."[49]

She becomes a link to Stahr's past, her physical similarities to his deceased wife, Minna, allowing Monroe to remain rooted to the traditional values of loyalty, trust, hard-work and morality he strongly believes in. His love and attraction for Kathleen represent more than a physical attraction, she symbolises everything that Stahr feels is slipping away from Los Angeles. The old values he holds in such high regard are eroding, pushed aside by the new materialistic values of money, power, and financial gain. Stahr's own physical deterioration coincides with the collapse of what he considers morally right in people. "The old loyalties were trembling now, there were clay feet everywhere; but still he was their man, the last of the princes."[50] When speaking about Hollywood and the film industry, F. Scott Fitzgerald commented:

> Hollywood is a strange conglomeration of a few excellent, over-tired men making the pictures, and as dismal a crowd of fakes and hacks at the bottom as you can imagine. [51]

Witnessing the film industry from such an intimate distance opens up the reality of the business to Fitzgerald, where the few good people are swallowed up by the nightmarish vision of a crowd of desperate people all trying to succeed in Hollywood.

If *The Last Tycoon'* centres around one of these "excellent men" in Monroe Stahr, then Nathanael West's *The Day of the Locust* is a portrayal of the crowd, the fakes, hacks and the disillusioned making up the majority of the Los Angeles population. With the entire country caught up in the Depression during 1930's, the city of Los Angeles personified the desperation of the American people in general and the attempt to mask the hopelessness of the situation. Kingsley Widner comments:

> West's insight was that the basic American repressed character was to merge with the Hollywood counterfeit – as it has in our puritanical decadence – providing the largest masquerade of civilization. Essentially, the historic Hollywood is dead; but just as essentially, America has become Hollywood. [52]

While the people of West's *The Day of the Locust* do not run film studios or have control over the content of the movies being made, they do represent West's belief that Los

Angeles was slipping towards an inevitable chaotic crisis. They are the spectators, the outsiders, influenced by other people's decisions, waiting for something to happen, unsure and unwilling to prevent the impending disaster, instead embracing it as an escape from their own ineffectual lives. Using the artist and set designer Tod Hackett's revelatory masterpiece, "The Burning of Los Angeles" in *The Day of the Locust*, West depicts Los Angeles and its people heading for a violent and uncontrollable finish. James F. Light comments:

> Having no life in themselves – no emotional vitality or beliefs or dreams – they must seek life elsewhere and often their search leads to the Sargasso Sea of dreams: Hollywood. Symbolically, then, these people – who exist everywhere – are suggested by the aged who have come to Hollywood to die physically the death they have always experienced emotionally. Bitterly they stare at the Hollywood scene. They feel cheated by California (and its paradisiacal promises) and cheated by life. [53]

Drifting to Hollywood aboard the promise of sunshine and success, dreamers arriving in Los Angeles realise the city cannot fill the void that controls their lives. Living in a city that promises so much but fails to deliver, the despairing multitude look for any escape from the mundane. They find one through viewing other people's lives, on film, in newspapers and in magazines, creating a fantasy world removed from the reality of their own existence. Tod Hackett, walking through the streets of Hollywood feels the eyes of the spectators on him. "When their stare was returned, their eyes filled with hatred." At this time, "Tod knew very little about them except that they had come to California to die."[54] Emotionally the spectators are already dead on the inside. They feel something is owed to them. "If only a plane would crash once in a while so that they could watch the passengers being consumed in a 'holocaust of flames' as the newspapers put it."[55] Satisfaction comes in the form of sensationalism, of incredible or devastating events that make the insignificant events of their own lives disappear from the memory.

Embodying this lack of emotional vitality but with other characteristics so lifeless

that he does not seem to exist at all is Homer Simpson. Coming from a small Mid-Western

town after a bout of pneumonia in order to enjoy the Californian sunshine, Tod comments

that, "this man seemed an exact model for the kind of person who comes to California to

die."[56] Homer's inability to express feeling, his awkwardness, insecurity and large

ineffectual hands are matched in absurdity only by the actions of the performers that

surround him. While Homer's internal repression is reflected by his outward persona and

the clumsiness of his movements, Harry Greener, another of the group, is a former

vaudeville performer who fails to acknowledge his own internal death, living his life as a

continuation of his act. Victor Comerchero comments:

> Like every other character in the novel, Harry must artificially stimulate feeling. To feel, he
> must act. This would be a grotesque ironic joke – the actor who acts so long that only the
> actor remains – if it were not for the darkness of the alternative.[57]

As a one-hit success of small notoriety, and a living relic of the silent era of film, Harry

refuses to confront reality, choosing instead to hide behind his comedy routine. "It was his

sole method of defence. Most people, he had discovered, won't go out of their way to

punish a clown."[58] Rather than submit to the defeat of an unfulfilled career, Harry chooses

to work out his days as a polish salesman, using the tired routine to sell his product. When

he becomes sick however, it becomes difficult to distinguish between his acting and the

reality of his pain. Acting, rather than accepting reality has become a popular form of

defence for many of the people living in Hollywood. The fancy dress and the landscape of

Los Angeles has become homogenised, unrecognizable as separate entities but a

continuous, flowing attempt at finding meaning in a meaningless life:

> Their sweaters, knickers, slacks, blue-flannel jackets with brass buttons were fancy dress.
> The fat lady in the yachting cap was going shopping, not boating; the man in the Norfolk
> jacket and Tyrolean hat was returning, not from a mountain but an insurance office; and the

girl in slacks and sneaks with a bandanna around her head had just left a switch-board, not a tennis court.[59]

The unreality comes not only in the way people dress but the houses they live in, their daily routine, the very attempt at trying to create a fantasy world of importance and relevance. Dead on the inside, their dreams unrealised, people try to create an outward appearance of happiness, contentment, success. David Fine comments:

> Identity is an utterly performative act, a matter of adopting screen poses. Characters walk
> the streets of Hollywood as if they were playing to the cameras, and the exotic houses they
> inhabit, aping and parodying every style in architectural history, look like they have been
> constructed on the movie lot by studio carpenters.[60]

There is no definitive line between where the dream ends and where reality kicks in. Unhappy in their 'real lives', people imitate what they perceive as a successful life, blurring the lines to a such an extent that even the imitation life is strikingly sad and desperate.

Hollywood becomes a symbol for deception; everything is and always will be a lie. Even socialising requires props. When Tod attends an industry party in *The Day of the Locust* he is led to a swimming pool to be shown the latest 'feature', a plastic reproduction of a horse lying on the bottom of the pool. When Tod points out it is merely a fake, he is chastised. "You're just an old meanie. Think how happy the Estees must feel showing it to people and listening to their merriment and their oh's and ah's of unconfined delight."[61] Nothing in Hollywood can be taken at face value; everything is manipulated into a machine for entertaining, to relieve the boredom of everyday life. Even a set mistake made on the lot of a film studio becomes paralleled to a famous historic event. While filming a scene about the famous Waterloo battle, a property manager sends a group of cavalry actors up an unfinished prop, causing it to collapse. "It was a classic mistake, Tod realised, the same one Napoleon had made."[62]

Tod too, is an outsider, a product of East Coast upbringing. Questioned by his fellow university graduates, who feel he has sold his soul to the corruption of Hollywood, Tod defends his position through the progression of "The Burning of Los Angeles", an interpretation of the impending chaos that will eventually grip the city. The surrealist tendency in the painters Hackett chooses as inspiration, Guardi and Desiderio, capture the illusion and falseness of the Hollywood landscape. Nothing is as it appears on the surface. The people of the city are desperately close to breaking point. As George M. Pisk comments:

> His painting is to represent the state of violence and anarchy when the locusts finally
> emerge from their dormant state. The people whom he intends to paint in this eruption are
> the same as those that appear on West's pages. [63]

The spectators and performers in West's *The Day of the Locust* eventually awaken from their dream-like slumber to become a violent mob, aroused by a common emotion generated by the forming of a faceless mass. They are not leaders, merely followers. After spending a lifetime inwardly alone, the crowd finds energy in sheer numbers. They are the same kind of people who attend Harry Greener's funeral. "While not torch-bearers themselves, they would run behind the fire and do a great deal of shouting. They had come to see Harry buried, hoping for a dramatic incident of some sort, hoping at least for one of the mourners to be led hysterically from the chapel."[64] The movie premiere at Kahn's Persian Theatre offers the backdrop for West's mob to come to life. Victor Comerchero comments:

> They are like a silent chorus in a Greek tragedy, writhing in the background. Their 'stare' is
> a mute protest, an accusation, and a commentary on the spiritual vacuum that is modern life.
> At the end of the novel, when they find voice, it is in blind, ruthless, consuming violence.[65]

The result of the crowd becoming a vicious, uncontrollable mob is a realization of Hackett's own artistic vision of the city burning. The original dream of Los Angeles as a city of sunshine and oranges, as a place for a new beginning or simply as a place to die

peacefully dissolves into a prophetic wasteland. The dark, violent streets of hard-boiled

literature spreads across the city, dismantling the bright face carefully put together by the

boosters. Life is shown to be temporary and fragile and the city as much dull or dangerous

as it is exciting and empowering. The falseness and emptiness of a landscape designed to

imitate paradise is exposed and then destroyed by the very people who had originally come

to the city as dreamers, only to be cheated and disillusioned by what they experienced.

---

[1] Blake Allmendinger, 'All About Eden', in *Reading California: Art, Image, and Identity, 1900-2000* ed. by Stephanie Barron, Sheri Bernstein, Ilene Susan Fort (Berkeley, CA: University of California Press, 2000), pp. 113-27.
[2] Mike Davis, *City of Quartz*, p. 3.
[3] Ibid, p. 4.
[4] James M. Cain, *The Postman Always Rings Twice*, in *Crime Novels – American Noir of the 1930's and 40's* (New York, NY: Viking Penguin, 1997), p. 3.
[5] Ibid, p. 7.
[6] David Madden, *James M. Cain* (New York, NY: Twayne Publishers, Inc, 1970), p. 108.
[7] James M. Cain, *The Postman Always Rings Twice*, in *Crime Novels – American Noir of the 1930's and 40's*, p. 22.
[8] T.H. Watkins, *The Great Depression – America in the 1930's* (Toronto, Canada: Little, Brown and Company, 1993), p. 195.
[9] Lee Horsley, *The Noir Thriller* (New York, NY: Palgrave Publishers, 2001), p. 18.
[10] June Howard, *Form and History in American Literary Naturalism* (Chapel Hill, NC: University of North Carolina Press, 1985), p. 48.
[11] , James M. Cain, *The Postman Always Rings Twice*, in *Crime Novels – American Noir of the 1930's and 40's*, p. 31.
[12] Ibid, p. 37.
[13] Paul Skenazy, 'Behind the Territory Ahead' in *Los Angeles in Fiction* ed. by David Fine (Albuquerque, NM: University of New Mexico Press, 1995),
[14] Ibid, p. 413.
[15] Gary J. Hausladen and Paul F. Starrs, 'L.A Noir', Journal of Cultural Geography, 23.1 (2005), p. 7, in *Academic Search Premier* <http://web.ebscohost.com.ezproxy.waikato.ac.nz:2048/ehost/pdf?vid=3&hid=20&sid=2e18eb40-81c5-4a80-b9dc-73c9dde70b45%40sessionmgr2> [accessed 16 February 2007].
[16] Joseph C. Porter, 'The End of the Trail: The American West of Dashiell Hammett and Raymond Chandler' in *The Western Historical Quarterly*, 6 (1975) p. 411, in *JSTOR* <http://links.jstor.org/sici?sici=0043-3810%28197510%296%3A4%3C411%3ATEOTTT%3E2.0.CO%3B2-1> [accessed 3 July 2006].
[17] Lee Horsley, *The Noir Thriller*, p. 25.
[18] Dashiell Hammett, *The Maltese Falcon* (New York, NY: Random House Inc., 1989), p. 3.
[19] Cynthia S. Hamilton, *The Western and Hard-boiled Detective Fiction in America* (Iowa City, IA: University of Iowa Press, 1987), p. 127.
[20] Frank McShane, *The Life of Raymond Chandler* (Toronto, Canada: Clark, Irwin and Company, Ltd., 1976), pp. 47-8.
[21] Colin Shindler, *Hollywood in Crisis – Cinema and American Society*, p. 63.

[22] Frank McShane, *The Life of Raymond Chandler*, p. 50.

[23] Raymond Chandler, *The Big Sleep* in *Three Novels* (Harmondsworth, UK: Penguin Books Ltd., 1993), p. 16.

[24] Ibid, p. 4.

[25] Megan E. Abbott, *The Street Was Mine* (New York, NY: Palgrave Publishers, 2002), p. 48.

[26] Raymond Chandler, *The Big Sleep*, p. 11.

[27] Ibid, p. 112.

[28] David Fine, 'Beginning in the Thirties: The Los Angeles Fiction of James M. Cain and Horace McCoy' in *Los Angeles in Fiction*, pp. 43-66.

[29] Horace McCoy, *They Shoot Horses Don't They?* in *Crime Novels – American Noir of the 1930's and 40's* (New York, NY: Viking Penguin, 1997), p. 112.

[30] Ibid, p. 113.

[31] David Fine, *Imagining Los Angeles*, p. 102.

[32] Horace McCoy, *They Shoot Horses Don't They?*, p. 133.

[33] David Fine, *Imagining Los Angeles*, p. 103.

[34] Horace McCoy, *They Shoot Horses Don't They?*, p. 143.

[35] Ibid, p. 138.

[36] Lee J. Richmond, 'A Time to Mourn and a Time to Dance: Horace McCoy's 'They Shoot Horses, Don't They?', *Twentieth Century Literature*, 17 (1971), p. 99, in *JSTOR* <http://links.jstor.org/sici?sici=0041-462X%28197104%2917%3A2%3C91%3AATTMAA%3E2.0.CO%3B2-9> [accessed 12 July 2006].

[37] Horace McCoy, *They Shoot Horses Don't They?*, p. 207.

[38] Ibid, p. 210.

[39] Raymond Chandler, *'The Notebooks of Raymond Chandler' and 'English Summer – A Gothic Romance'* (London, UK: Weidenfeld and Nicolson, 1977), pp. 66-67.

[40] Christopher Ames, 'Shakespeare's Grave; The British Fiction of Hollywood', *Twentieth Century Literature*, 47 (2001), p. 409, in *JSTOR* <http://links.jstor.org/sici?sici=0041-462X%28200123%2947%3A3%3C407%3ASGTBFO%3E2.0.CO%3B2-O> [accessed 7 February 2007].

[41] K. G. W. Cross, *Scott Fitzgerald* (Edinburgh, Scotland: Oliver and Boyd, 1964), p. 108.

[42] F. Scott Fitzgerald, *The Last Tycoon* (Harmondsworth, UK: Penguin Books Ltd., 1977), p. 56.

[43] Ibid, p. 59.

[44] Ibid, p. 113.

[45] Ibid, p. 116.

[46] John Aldridge, 'Fitzgerald: The Horror and the Vision of Paradise' in *'F. Scott Fitzgerald: A Collection of Critical Essays'* ed. by Arthur Mizener (Englewood Cliffs, NJ: Prentice-Hall Inc., 1963), p. 108.

[47] F. Scott Fitzgerald, *The Last Tycoon*, p. 32.

[48] Ibid, p. 32.

[49] Ibid, p. 34.

[50] Ibid, p. 34.

[51] F.S. Fitzgerald, 'A Confluence of Voices I', in *Nathanael West: The Cheaters and the Cheated* ed. by David Madden (Deland, Florida: Everett/Edwards Inc, 1973), p. 173.

[52] Kingsley Widmer, 'The Last Masquerade', in *Nathaniel West: The Cheaters and the Cheated* ed. by David Madden (Deland, Florida: Everett/Edwards Inc, 1973), p. 177.

[53] James F. Light, 'Nathanael West and the Ravaging Locust', *American Quarterly*, 12.1 (1960), p. 47, in JSTOR <http://links.jstor.org/sici?sici=0003-0678%28196021%2912%3A1%3C44%3ANWATRL%3E2.0.CO%3B2-D> [accessed 19 July 2006].

[54] Nathanael West, *The Day of the Locust* (Middlesex, UK: Penguin Books Ltd, 1963), p. 10.

[55] Ibid, p. 148.

[56] Ibid, p. 32.

[57] Victor Comerchero, *Nathaniel West: The Ironic Prophet* (Syracuse, NY: Syracuse University, 1967), p. 139.

[58] Nathanael West, *The Day of the Locust*, p. 30.

[59] Ibid, p. 10.

[60] David Fine, *Imagining Los Angeles*, p. 155.

[61] Nathanael West, *The Day of the Locust*, p. 23.

[62] Ibid, p. 97.

[63] George M. Pisk, 'The Graveyard of Dreams: A Study of Nathaniel West's Last Novel, "The Day of the Locust"', *The South Central Bulletin*, 27.4 (1967), p. 65, in JSTOR <http://links.jstor.org/sici?sici=0038-321X%28196724%2927%3A4%3C64%3ATGODAS%3E2.0.CO%3B2-7> [accessed 20 July 2006].

---

[64] Nathanael West, *The Day of the Locust*, p. 89.
[65] Victor Comerchero, *Nathaniel West: The Ironic Prophet*, p. 148.

CHAPTER TWO

In the early hours of December 7th, 1941, the Japanese Empire launched a full-scale offensive against the United States, attacking the strategic naval location of Pearl Harbour in the Hawaiian Islands. America was at war. With an immediate need for munitions and military equipment, the nation, still struggling to overcome the effects of the Depression, was able to provide a workforce to meet the demands of war. In Los Angeles, the aeronautic industry benefited from World War II and the need for workers. Migrants, on an unprecedented scale came to Los Angeles from 1940 onwards into what Edward Soja describes as "mass suburbanization on a scale never before encountered."[1] Rejuvenated with a sense of opportunity and patriotic duty, the people of Los Angeles once again began to prosper, from protecting and serving their country. Hollywood too played a part in assisting a growing war machine. Stars like James Stewart, Clark Gable and Robert Montgomery contributed to the nation's cause and "ordinary doughboys could be seen dancing with Joan Crawford or being served food by Katherine Cornell: proof that the stars yielded to no one in their allegiance to democracy."[2] On the surface the anxiety and desperation caused by the Depression appeared to have passed, but as the city of Los Angeles continued its rapid expansion, cracks appeared in the picture of unity.

The Second World War changed Los Angeles and it changed its literature. With industrial growth and ongoing suburban sprawl, the physical and cultural identity of Los Angeles became one of diversity and separation, a direct contrast to the picture of unity displayed by Hollywood , and an alien concept to the compact nature of Eastern cities like New York and Boston. Combining the vastness and disconnectedness of the city with the superficial nature of architecture influenced by a film industry that turned street corners into shooting sets, Los Angeles became a confusing, dysfunctional, metropolitan area. This was vividly reflected by the people who populated the city and the people writing about it. The city was no longer a recognizable as a single entity but a jumble of economic, cultural and social outposts with names like Fontana, Chinatown and Beverly Hills. The landscape and architecture of Los Angeles that were satirically portrayed in Tod Hackett's painting, 'The Burning of Los Angeles' in Nathanael West's *The Day of the Locust* seemed a reality.

A series of novels published after the war demonstrated that, despite the boom generated by war, industry and immigration, the city remained one of promise unfulfilled. Evelyn Waugh's *The Loved One* (1948) shows an outsider's view of Los Angeles, replete with British superiority. The novel highlights the alienation of the outsider, characters are emotionally detached and the city tries too hard to create a past where none exists. Norman Mailer's *The Deer Park* (1955) tells the story of three men, a film director, an air force veteran who finds no place in America after the war, and a pimp, both ambitious and philosophical, who serves as a modern Faust. Director and screenwriter Charles Eitel is blacklisted from Hollywood after refusing to answer the questions of McCarthy's Congressional Committee on Un-American Activities. After Charles takes exile in Desert D'or, a desert town two hundred miles from the cinematic capital, he is faced with the stark reality that once removed from his artistic work, his life is meaningless, simply a series of empty, shallow relationships and conversations. Allen Ginsberg's *Howl* (1956) and

Mailer's 'The White Negro' (1957) challenge the American political environment during

the emergence of the Cold War that followed World War II. They voice their

disillusionment with the bleak future created by extreme anti-communist committees and

the use of nuclear weapons. Thomas Pynchon's *The Crying of Lot 49* (1966) reverses the

earlier model of the California detective story portrayed by writers like Raymond Chandler

and Dashiell Hammett. The quest undertaken in *The Crying of Lot 49* by the protagonist

Oedipa Maas is an attempt to find links between the contemporary world she lives in and a

meaningful past. Her quest takes place, however, against a backdrop of entropy that shows

a closed world losing energy as time decays and order disintegrates toward chaos. Finally,

Joan Didion's *Play It As It Lays* (1970) shows Maria Wyeth using the Los Angeles freeway

as her escape from alienation. Comforted by the familiarity of exit ramps, billboards and

the continuity of driving, Maria searches for meaning in a society dominated by emotional

emptiness, identity loss and contradiction.

Separated by more than a decade, the future in Evelyn Waugh's *The Loved One* is

every part as desperate as Thomas Pynchon's final auction in *The Crying of Lot 49*, the

continuation of life is unsettled and unknown. The principal setting of the novel is the

Whispering Glades, a funeral park in Los Angeles created "to elevate the Soul of Man"[3]

and to give people "the certain knowledge that their Loved Ones were very near, in Beauty

and Happiness."[4] The main objective of the staff at Whispering Glades is to make people

aware of their inevitable death while they are still living. When Dennis Barlow first goes to

Whispering Glades to arrange Sir Francis' funeral, he is informed, "many of our friends

like to make Before Need Arrangements."[5] To maintain an image of being well-loved and

thought of, even in death, is an important part of overall appearance. The proprietors of

Whispering Glades recognise that importance, and charge exorbitant fees to create a false

sense of peace for a city of people fixated on their image and waiting to die.

For Waugh, the death of the American dream at the hands of Los Angeles symtomises the inertia of Western civilisation. Using the replicated architecture of some of Europe's most recognizable cultural and religious landmarks, the funeral park of Whispering Glades attempts to find a historical past in a city that does not have one. Walter Wells comments:

> The founder's dream, however, is not to create new cultural artifact, new life, but rather to plunder, through imitation, the treasures of other cultures so as to vitiate death. The dream, which in earlier Hollywood novels had been American and dying, became to Waugh the living death of western civilization.[6]

From the perfect replica of an English Manor to a marble replica of Rodin's statue, 'The Kiss', all of the attempts to stylize and attach death to art cannot cover up the fact Los Angeles is a city populated by individuals lost, lonely and looking for meaning while they are still alive. Those Loved Ones waiting for "inhumement, entombment, inurnment or immurement"[7] at Whispering Glades are no different from the unidentified homeless man who dies in his sleep on an unmarked park bench. From the very beginning of *The Loved One*, Los Angeles is compared with a darkness normally associated with the wilderness of the jungle or desert. The two Englishmen who are introduced while sipping drinks on a porch become merely, "the counterparts of numberless fellow-countrymen exiled in the barbarous regions of the world."[8] Los Angeles, as seen through Waugh's eyes, is no more civilised than any of the more untouched parts of the world.

Waugh plays on the fact many Los Angeles residents are themselves outsiders, coming from small towns, eastern cities, or different countries in search of success, and are never accepted or feel comfortable. The character of Sir Ambrose Abercrombie is one such person who, after coming from England to play certain stereotypical 'English' roles, maintains his standing in Hollywood society through role-playing these same characters. During the hot Los Angeles summer, he wore, "dark grey flannels, an Eton Rambler tie, an

I Zingari ribbon in his boater hat," and, "whenever the weather allowed it he wore a deer-stalker cap and an Inverness cape."[9] Abercrombie is desperate to maintain a proper British image, a unified front that celebrates the closeness of the English community. When Sir Francis Hinsley is released from his writing job at a Hollywood studio and subsequently hangs himself, Abercrombie comments, "Everything depends on reputation – "face" as they say out East. Lose that and you lose everything. Frank lost face. I will say no more."[10] What Abercrombie fails to realise, or is too afraid to admit, is how reliant he has become on his imagined persona in order to survive in Los Angeles. Without the façade of his very English gentleman antics, Abercrombie would be destined for the scrapheap of tired routines. Much like the retired vaudeville actor Harry Greener in *The Day of the Locust*, Abercrombie's daily act is a matter of self-preservation, the only way he remembers how to exist.

Falseness and illusion are ever-present in *The Loved One*. From Dennis Barlow's attempt to keep his work at The Happier Hunting Grounds secret, to Mr Joyboy's double life as a respected cosmetician and repressed son, no-one is who they appear to be. Yet Waugh's impression of Los Angeles, told through the eyes of Barlow, make him the ultimate outsider, never comfortable in the strange city and emotionally indifferent to the people around him. In the love triangle involving Barlow, Joyboy and Aimee Thanatogenos, Barlow presents himself as a poet to Aimee, winning her admiration and replacing, "that impersonal insensitive friendliness which takes the place of ceremony in that land of waifs and strays."[11] Aimee can only relate to Barlow when she associates him with the beauty of what she believes to be his writing. When she discovers the truth about Barlow's occupation and his plagiarised poetry, she has become attached to him and finds herself unable to end the relationship for fear of seeming a poor judge of character. "The exposure as a liar and a cheat of the man she loved and to whom she was bound by the tenderest vows, affected only a part of her."[12]

A product of the Los Angeles obsession with appearance, Aimee maintains her bond with Barlow, but is thrown into further turmoil by her relationship with Mr Joyboy, a man she deeply admires as a cosmetician. Her admiration is tainted when the relationship is taken outside the workplace. Joyboy's house is "not a prepossessing quarter, it seemed to have suffered a reverse,"[13] and Aimee has her first realisation of Joyboy's imperfections. Confused and without the aid of a confidante, Aimee turns to a newspaper advice column for support. Like the writer in Nathanael West's *Miss Lonelyhearts*, the Guru Brahmin in *The Loved One* is a disillusioned male, emotionally drained from hearing about the uncountable acts of sadness occurring in people's lives everyday. The Guru's final piece of advice to Aimee is simple. He tells her that the only escape is through death. Confounded by this advice and unable to break a lover's vow perceived to be sacred, Aimee takes the Guru's words literally.

If the people in *The Loved One* are false, and driven to endlessly replicate the commercially generated environment in which they live, Aimee presents herself to the world, "dressed and scented in obedience to the advertisements."[14] The architecture of the novel also exposes Los Angeles as a confused city. Like the food markets in West's *The Day of The Locust*, where "the oranges were bathed in red, the lemons in yellow, the fish in pale green,[15] the landscape in *The Loved One* is never in synch with the outside world, something that hints of the stage set. Whispering Glades, a place to connect with a past, either real or imagined; an opportunity to feel accepted and remembered, is really neither. Fredrick Beaty comments:

> Having no history, no roots, and no culture they can call their own, they are, like birds of passage, constantly looking back toward their true home somewhere else.[16]

The Pilgrim's Nest, Lover's Nest and Poet's Corner all offer a final resting place where in death, unlike in life, the individual seems to belong somewhere, and are surrounded by others to whom they seem connected. The influence of Hollywood is also present at Whispering Glades with a special section, Shadowland, reserved for people involved in the film industry. While Shadowland is an exclusive site in the cemetery, the word shadow reveals the nature of the film industry. It is impossible to see anything in the full light and something is always hidden. The architecture, the replicated artwork and the people all reflect the falseness of a "land of sunshine and oranges placed on the cultural extremity at which the more extravagant form of American dream and American fantasy are made manifest, not only on the stage-sets and in the studios of Hollywood, but in the surrounding environment."[17]

In Norman Mailer's *The Deer Park*, the "confusion between reality and illusion, the conflict between commerce and art, the commodification of sex, and the erasure of both identity boundaries and an ordered sense of time and place,"[18] are explored. The novel reveals the Hollywood dream as false and pretentious. Following narrator Sergius O'Shaugnessy, a discharged air force pilot, and Charles Eitel, a blacklisted Hollywood director, *The Deer Park* reveals the existence of an exclusive society in the desert retreat of Desert D'Or. What on the surface appears to be a meeting place for the rich and famous is nothing more than a renamed former mining town posing as an architecturally fraudulent escape for the bored and lost people of Hollywood.

When Sergius first arrives in Desert D'Or with fourteen thousand dollars courtesy of a poker game, he is immediately taken by the confusion of time caused by the architecture. Sergius muses:

> Drinking in that atmosphere, I never knew whether it was night or day, and I think that kind of uncertainty got into everybody's conversation. Men lacquered with liquor talked to other men who were sober, stories were started and never finished.[19]

The bars, clubs and lounges, made to look like, "a jungle, an under water grotto, or the lounge of a modern movie theatre,"[20] shut people away from reality, creating the illusion of timelessness while the world continues to progress outside the tinted windows and smoky ceilings. It is this kind of escape Sergius relies on. Sickened by the growing realisation of atrocities caused by war and pilots like himself, Sergius believes, "it was better not even to think of this." He liked "the other world in which almost everybody lived. The imaginary world."[21] Attaching himself to the Hollywood crowd, who are eager to forget the past, Sergius believes he can begin to forget about his own past as a pilot, and as an orphan. What *The Deer Park* exposes in the town of Desert D'Or is that there is no where to get away from the horrors of real life, people are willing to sell their soul and mind to the highest bidder and moral integrity counts for little.

Charles Eitel begins his journey trying to escape from the hypocrisy of Hollywood but is eventually sucked back into the system. Blacklisted by the film studios after failing to co-operate with a committee interested in exposing communist activity in Hollywood, Eitel becomes a high profile victim like many others in Los Angeles during the Cold War struggle beginning after World War II. Suspicious and insecure, political groups like the House Committee for Un-American Activities signified the growing fear by many Americans of the 'evils' of communism infiltrating society. Eitel, unable to comprehend the severity of his situation makes a moral stand against the committee and finds himself exiled. Faced by the reality he may never make films again, Eitel is forced to soften his moral standing, questioning his decision not to co-operate. Theorising about his work, Eitel "saw now that he had one true love – those films which had flowered in his mind and never been made."[22] In betraying his true passion, he has betrayed himself. Eitel, like Sergius is desperate to forget his past, the past "a cancer, destroying memory, destroying the present, until emotion was eroded and the events in which one found oneself were always in danger of being as dead as the past."[23] Trying to outrun his past, Eitel succeeds

only in returning to his beginnings, caught in the circular nature of life in Los Angeles,

symbolised in earlier writing by Horace McCoy's marathon dance competition. Attempting

to make a clean start, Eitel sells a working script to Collie Munshin, a high profile

executive, who like Sammy Glick in Budd Schulberg's *What Makes Sammy Run?* has

bullied and slept his way to the top of the Hollywood ladder. Eitel, after making the deal

with Munshin realises he has returned to where he never wanted to be, "as one of the peons

Collie keeps locked in a hole."[24] Trapped by the studio's demands, Eitel's vision has been

corrupted, a position shared by many writers who came to Los Angeles with the dream of

gaining financial and artistic freedom.

Sex and relationships too become commodities in the search for appearance.

Sergius has a relationship with screen siren Lulu Meyers, "possessing, even desecrating, the

erotic fantasies of millions of moviegoers,"[25] but the reality is that the love affair holds no

future and is only temporary. Lulu may have some love for Sergius but her dream relies

too much on her appearance, she is always acting and Sergius says, "it took me longer than

it need have taken to realize the heart of her pleasure was to show herself."[26] Hollywood

feeds off positive images and Lulu gains satisfaction not in her personal relationship with

Sergius but with the popularity of the public's perception of her. Sergius eventually sees

the role he plays for Lulu, as the orphan, pilot prince, an accessory to compliment the

image of a movie star and win the admiration of the public. On the other hand, Eitel's

relationship with Elena Esposito, the cast-off mistress of Munshin mirrors the actions of his

working life in Los Angeles. By courting Elena, Eitel has made a conscientious decision to

shun his normal Hollywood-type partner, someone only interested in how others view

them, and finds Elena revitalising and different. Yet, as he discovers his urge to make movies a necessity to continue living, Elena is turned into an illusion, a false image of something that cannot exist in reality. Eitel comments:

> The unspoken purpose of freedom was to find love, yet when love was found one could only desire freedom again. So it was. He had always seen it as a search. One went on, one passed from affair to affair, some good, some not and each provided in its own way a promise of what could finally be found. How sad to finish the journey and discover that one was unchanged, that indeed one was worse; still another illusion was lost.[27]

Elena represents Eitel's yearning for something better from life, but in returning to testify in front of the same committee that had earlier ostracized him, he chooses his love of film-making over the love of another human being.

Marion Faye, son of socialite Dorothea O'Faye, encompasses everything Hollywood tries to hide from. He is the product of a passing affair between a beautiful nightclub singer, Dorothea, and a European prince. The union is a romantic Hollywood notion, but in reality, Marion's birth is a mistake. Rejecting the false, chaste image that Hollywood projects to the world, Faye becomes a pimp. Mark Winchell comments:

> He is presented as a kind of existential saint who carries his war against hypocrisy to perverse lengths. As such, he is the antithesis of Hollywood. Hollywood preaches chastity while practicing debauchery. Faye is quite openly a pimp. Hollywood projects an image of compassion and humanity while actually destroying the lives and talent of its most creative people; Faye gratuitously torments a heroin addict and deliberately seeks to drive his mistress to suicide.[28]

While Charles Eitel retreats from Los Angeles and Hollywood momentarily before succumbing and returning, and Sergius is almost enticed into the world of movies, Faye remains determined to rebel against the falseness of the industry. Faye, disillusioned by Hollywood, refuses to be caught up with the romantic notions of success associated with the film world. Driving through the desert, Faye is overwhelmed by the desperate nature of society, begging for an atomic bomb, "to clear the rot and the stench and the stink," to "let it come for all of everywhere, just so it comes and the world stands clear in the white dead dawn."[29]   Like Waugh, he views Hollywood as symptomatic of the disintegration of society. He cannot understand why people continue to lead meaningless, painful lives. Where Robert Syverton makes a gradual journey towards understanding Gloria's nihilistic vision of the future in *They Shoot Horses Don't They?*, Faye constantly pushes Elena into taking her own life to escape. The illusion Hollywood presents never confronts harsh reality of life but glosses over the memories of a tragic, confusing past with bright, hopeful images for a successful future.

The conclusion of World War II, the realisation of atrocities caused during the Holocaust and atomic bombings of Hiroshima and Nagasaki, alongside the potential for nuclear engagement, engendered alienation from society among the young, and spurred a multitude of artistic responses. Ginsberg's 'Howl' came to epitomise the disillusion of a beaten generation. "I saw the best minds of my generation destroyed by madness, starving hysterical naked,…who drove cross country seventy-two hours to find out if I had a vision or you had a vision or he had a vision to find out eternity."[30] Disillusioned by what they believed was an inevitably apocalyptic future, along with the repressive nature of a paranoid Government, Beat writers Allen Ginsberg, William S. Burroughs and Jack Kerouac attempted to fill the artistic vacuum with controversial but aesthetic and intellectual discoveries. John Tytell comments:

> Beginning in despair, the Beat vision was elevated throughout the shocks of experience to a realization of what was most perilous about American culture.[31]

The writers of the Beat generation attempted to overcome the stifling atmosphere of Cold War America through highlighting and attacking the ideas threatening to extinguish individuality and creative spirit. Ginsberg's 'Howl' resonated from San Francisco when its

publisher was tried and acquitted for issuing that allegedly obscene work. It was this trial
that made the poem famous.

The participation of Los Angeles' images of alienation and despair, in a more
general American disquiet can be demonstrated by a consideration of Norman Mailer's
essay 'The White Negro'. Speaking about the Cold War, Mailer comments:

> ...the atmosphere that lay over America then was much more directed and overbearing and
>
> overpowering and disagreeable. It was not that we were in any way a fascistic state, except
>
> in one serious fashion, which is that people were beginning to be ready to not think freely.[32]

With the affects of an atomic war and a desolate outlook for the future creating new schools
of thought, people that had been previously subdued and frightened into popular thinking
began to voice original and controversial ideas. Mailer comments:

> One could hardly maintain the courage to be individuals to speak with one's own voice, for
>
> the years in which one could complacently accept oneself as part of an elite by being a
>
> radical were forever gone. A man knew that when he dissented, he gave a note upon his life
>
> which could be called in any year of overt crisis. No wonder that these have been the years
>
> of conformity and depression. A stench of fear has come out of every pore in American
>
> life, and we suffer from a collective failure of nerve. The only courage, with rare
>
> exceptions, that we have been witness to, has been the isolated courage of isolated people.[33]

The ability to exercise the right to freedom of speech had been lost in a time when
uncertainty was the over-riding feeling for many people. The fear hanging over the nation
like an oppressive cloud preventing individuality and creative freedom from blossoming, is
perhaps intensified in the literature of Los Angeles. 'The White Negro' is regarded as a
white male investment in African American masculinity and the idea of the hipster as
individually alternative and sexually provocative. The confrontation in "The White Negro'
becomes a choice, "one is a rebel or one conforms, one is a frontiersman of the American
night life, or else a square cell."[34] The individual, like the long mistreated African
American "meets generally so much enmity, competition, and hatred in the world of Hip,

that his isolation is always in danger of turning upon itself, and leaving him indeed just

that, crazy."[35] The journey of the Hipster, of the individual ready to confront his repressed

identity, is to discover meaning and voice an opinion, without ever losing his sense of

reality.

Ginsberg's *'Howl'* is the journey of a hipster travelling an emotional wasteland,

populated by disillusioned, half-crazed shadows of living people. The promises that had

once made Los Angeles the final frontier for opportunity and success had evaporated. The

people, "who wandered around and around at midnight in the railroad yard wondering

where to go"[36] had been abandoned. The despair and desperation resonating throughout

'Howl' reveals no utopian destination. A product of a dysfunctional family and obsessed

by the absolute power of the nuclear world, Ginsberg uses *'Howl'* to commentate on a

complex network of social and personal conflicts. Jonah Raskin comments:

> *'Howl'* emerged from the fissures in American society after World War II, as well as from
>
> the fault lines in the author's own secret, volatile life. In his best poetry, Ginsberg dove into
>
> the wreck of himself and of the world around him to salvage himself and something worth
>
> saving of the world at large. In the act of writing *'Howl'*, he discovered the very language
>
> he needed – a language of the everyday and of Judgment Day – a language of the mundane
>
> and the apocalyptic.[37]

In his attempt to break free from the political and cultural shackles restraining society,

Ginsberg presents a picture of a nightmare world. The tragedy and persecution he writes

reflect the fears and disappointments, not only of his generation, but of earlier generations

of seekers like the long line of dreamers that came to Los Angeles looking for redemption.

Ginsberg has seen how society can transform and destroy an individual. In the second

section of *'Howl'*, the philistine God, Moloch, is invoked to symbolize the commercialism,

industrialization, sexual repression and death of spirit that has invaded the American

landscape. Gregory Stephenson comments:

> American society is seen as having consistently ignored, suppressed, and destroyed any
> manifestation of the miraculous, the ecstatic, the sacred and the epiphonous.[38]

The post-atomic environment, aided by the increasing dominance of American industry was threatening to destroy the soul and spirit of the American people with all of their, "Visions! omens! hallucinations! miracles! ecstasies! gone down the American river!"[39] and lost forever. With Los Angeles at the forefront of aeronautic and munitions production during and after World War II, the city became guilty of destroying the pure American Dream, accelerating the apocalyptic vision of a nuclear future.

Los Angeles, with its explosion of suburban sprawl, technological advancements in the aeronautic-space industry and with the House Committee of Un-American Activities breathing heavily over Hollywood's shoulder, did not encourage people to think freely. Ginsberg's 'Howl' is filled with the agony and madness of a society on the brink of destruction, but the poem's final images offer the possibility of a spiritual revival to combat the emptiness. Stephenson comments:

> Confinement, repression, alienation, and the dark night of the soul are ended. The
> "imaginary walls collapse" (walls of egotism, competition, materialism – all the woes and
> weaknesses engendered by Mental Moloch), and the human spirit emerges in victory,
> virtue, mercy and freedom.[40]

If the world has been turned into a dark place by the machines of war and industrialization, it can be reformed into a state of enlightenment and fulfilment through personal understanding. It becomes the individual's right to perceive and envision a brighter future wherever they live, "Who digs Los Angeles IS Los Angeles"[41] If Los Angeles has been sucked under by the greed and narrow-mindedness of financial gain, it can also be redeemed. The same people forced into spiritual isolation can rebel against the system and change the outlook for future generations.

The spiritual crusade against a desolate future is replaced by confusion in Pynchon's *The Crying of Lot 49*. Unlike the earlier detective novels of Raymond Chandler and Dashiell Hammett, where the object was to take a series of complicated situations and simplify them, *The Crying of Lot 49* works in reverse. As Edward Mendelson comments:

> *The Crying of Lot 49* starts with a relatively simple situation, and then lets it get out of the heroine's control: the simple becomes complex, responsibility becomes not isolated but universal, the guilty locus turns out to be everywhere, and individual clues are unimportant because either clues nor deduction can lead to the solution.[42]

The world that Oedipa Maas lives in and has previously understood becomes corrupted when she is given the responsibility of executing her former lover, Pierce Inverarity's, estate. This seemingly straightforward event becomes more and more difficult as she uncovers the possibility of an underground, anarchic organization called the Tristero, with origins in 13th century Europe, now operating in California. As Oedipa discovers further clues she is unable to determine if the Tristero organization is real, if she is part of an elaborate hoax arranged by Inverarity, or if she is simply hallucinating. As Geoffrey Lord comments:

> Oedipa's experience can be seen in part as a "revelation" – her discovery – of this world, the concrete realization of "America" in which she can find no sense of order and place. The novel's location seems to lack substantiality, and Pynchon's America seems to be fantastic.[43]

The implausibility of a centuries-old renegade mail organization, the disconnectedness of the people and places surrounding Oedipa, and the impossibility of finding concrete evidence lead her to question her own sanity, and the sanity of the people around her. The ambiguous nature of historical fact leads Oedipa to uncover more questions than answers. Terry Eagleton comments:

> What postmodernism refuses is not history but History – the idea that there is an entity
> called History possessed of an immanent meaning and purpose which is stealthily unfolding
> around us even as we speak.[44]

The fictional city of San Narciso, like Los Angeles, is built around a collection of images,

not concrete truth. The whole world has become Los Angeles, where meaning and purpose

may be dissolved and re-constituted as something else. While the boosters of Los Angeles

promoted a fabricated view of sunshine and healthy living to increase growth, San Narciso

is founded and grows on the underlying suspicion of a Cold War nuclear race.

As Oedipa drives on familiar freeways to San Narciso, it seems even the buildings

and places around her are trying to communicate in code, the "vast sprawl of houses that

had grown up together"[45], reminding her of a radio's circuit card, the pattern a

"hieroglyphic sense of concealed meaning, of an intent to communicate."[46] Driving the

freeway becomes synonymous with being caught in a maze. The city, a labyrinth of

buildings and places joined together in a haphazard manner, is linked only by the endless,

circular route of the road. Communication is homogenized, everything is at once connected

and separated; society, media and technology have made it impossible to recognize

anything as a separate entity. The individual's reliance on media and technology is

highlighted by the former actor turned lawyer, Metzger. Patricia Burgh comments:

> It appears a fact of life that real life is analogous to reel life. People assess their
> professional, academic, parenting, sexual and recreational patterns upon images provided by
> TV and movie characters.[47]

Hollywood studios and television networks have programmed people to believe that media

images portray real life. Metzger uses his role of former child actor to seduce a willing

Oedipa in a game of strip-questioning as they watch a re-run of an old film he has starred

in. Dressed in all the clothing she has in her suitcase and becoming increasingly drunk,

Oedipa loses herself in a maze of images, "the succession of film fragments on the tube, the progressive removal of clothing that seemed to bring her no nearer nudity, the boozing, the tireless shivaree of voices and guitars from out by the pool."[48] The line between fiction and real-life blurred when "she went into the bathroom, tried to find her image on the mirror and couldn't."[49] The constant outpouring of information through real and invented sources make it difficult to maintain a conscious understanding between what is fiction and reality. Even physical pleasure is entwined with technology and media, "her climax and Metzger's, when it came, coincided with every light in the place, including the TV tube, suddenly going out, dead, black."[50]

The journey Oedipa undertakes to uncover the 'truth' about Tristero becomes both an attempt to find a meaningful link between the past and the present, and a discovery of self. Lord comments:

> As clue after clue "turns up," Oedipa becomes increasingly absorbed in the desperate pursuit of either some definite confirmation, or – just as important – confutation, of the existence of Tristero. But nothing she discovers is verifiable; everything presents an appearance of artifice, including seemingly actual facts, events and people.[51]

The line between reality and unreality blurs as each meeting or event leads to a new, confusing situation. Each person Oedipa meets reveals new cryptic information, taking her further from resolution. The chance meeting at a bar has Oedipa introduced to Mike Fallopian who reveals the Peter Pinguid Society to her, an organization that believes the Cold War actually started in 1864. The historical facts in which Oedipa put her faith, dates, times and people, are dismantled. Tony Tanner comments:

> Oedipa's problem is whether she has in fact discovered, stumbled across, been lured by, a genuinely alternative mode of communication which does convey real messages in a way that subverts the 'government monopoly' – namely, the Tristero (another name that invites being played with – a meeting with sadness and terror, for example). And if so, will it release her from her isolation – or confirm it?[52]

Like the Tristero plot before it, whether or not a confrontation ever took place is

inconsequential. The mere possibility of the event taking place meant, "the ripples from

those two splashes spread, and grew, and today engulf us all."[53] The members of the Peter

Pinguid Society are only interested in the destructive nature of the confrontation and the

possible connection to a nuclear future. In common with others who came to California

and Los Angeles in particular, they see the evil in both industrial capitalism and Marxism

because, "underneath, both are part of the same creeping horror."[54] Even the revered

defender of American soil, Peter Pinguid, is seduced by the American dream of success;

after the run-in with the Russians he spends the rest of his life growing wealthy on

California real estate speculation.

Escape from the labyrinth of unanswerable questions, and the futility in finding

resolution to problems, is handled in extreme measures throughout *The Crying of Lot 49*.

The people Oedipa relies on for support and guidance desert her. She comments:

> My shrink, pursued by Israelis, has gone mad; my husband on LSD, gropes like a child
> further and further into the rooms and less rooms of the elaborate candy house of himself
> and away, hopelessly away, from what has passed, I was hoping forever, for love; my one
> extra-marital fella has eloped with a depraved 15-year-old; my best guide back to the
> Trystero has taken a Brody.[55]

Mucho Maas, Oedipa's husband, haunted by his past occupation as a second-hand car

salesman has turned to LSD, effectively turning the repetitive nightmare warning of a car

sign into what it actually is, a simple, creaking, metal sign. "Now he would never be

spooked again, not as long as he had the pills."[56] By taking the pills, Mucho escapes his

fears but alienates himself from Oedipa. Alone and confused, even Oedipa's conviction,

the "unvoiced idea that no matter what you did to its edges the true Pacific stayed inviolate

and integrated"[57] has been shattered by Randy Dribblette's decision to walk into the Pacific

Ocean. There is nowhere to go. The Pacific Ocean is the end of the road and with it, the end of the dream. Everything that would point to evidence is ambiguous, Oedipa cannot be

certain if she has stumbled, "onto a network by which X number of Americans are truly

communicating whilst reserving their lies," or if she is, "hallucinating it. Or a plot has been

mounted against you," or, "you are fantasying some such plot, in which case you are a nut,

Oedipa, out of your skull."[58] While Ginsberg's 'Howl' vented frustration and despair out

loud in a revitalising, soul-cleansing shout, Oedipa Maas stews in her uncertainty, trusting

nothing. Believing less and losing sight of the difference between reality and fiction,

Oedipa is a modernist heroine trapped in a post-modern world, searching for a master

narrative where none may exist.

Central to Oedipa's quest is her search for concrete evidence on W.A.S.T.E, the

American branch of the rebellious Tristero faction who were trying to undermine the

national mail service. Richard Lehan comments:

> They are intent in their belief that machines cannot go on forever, leaving waste by-
> products; they also seem to believe that the communication and information systems
> (including the mail) through which power is sustained, can be undermined, leaving silence.
> Out of the entropic waste and silence comes the death of the System, and once again we
> have the suggestion of an apocalypse.[59]

The apocalyptic vision of the Los Angeles mob burning the city to the ground in Nathanael

West's *The Day of the Locust* is transformed into a silent organization of disgruntled people

waging a sophisticated war against the Government machine of capitalism through an

underground mail system. Pynchon's W.A.S.T.E movement believes that by abandoning

traditional forms of communication and creating a message vacuum, the ensuing silence

will give power back to the people. The outcome then, is a shared belief, that with the

death of the Government, or an acceptance on their part that they abused the system, the

power will be returned to the people. The impossibility of ever knowing whether or not the

W.A.S.T.E movement exists remains with Oedipa until the very end, and the foreseeable future is questionable and frightening.

If the journey of the Hipster is to find a voice, and the journey of Oedipa Maas in *The Crying of Lot 49* is an attempt to find answers and meaning in the past, Maria Wyeth's quest in Joan Didion's, *Play It As It Lays* is to forget her past, to erase the painful memories of her existence. Like Horace McCoy's Gloria in *They Shoot Horses Don't They?* Maria has a nihilistic view on life; she finds no reason to be alive but is unwilling to die, viewing death as an easy escape from the pain and despair of living. Maria attempts to erase the past through her daily journeys on the Los Angeles freeway. While, "at night the dread would overtake her, bathe her in cold sweat, flood her mind with sharp flash images," once driving her despair disappears, if only for a short time as, "she never thought about that on the freeway."[60] Didion, in her essay 'Bureaucrats' says of driving in Los Angeles:

> Actual participation requires a total surrender, a concentration so intense as to seem a kind of narcosis, a rapture-of-the-freeway. The mind goes clean. The rhythm takes over. A distortion of time occurs, the same distortion that characterizes the instant before an accident.[61]

In focusing all of her attention on driving, Maria can forget everything that has happened in her life and everything that will happen in the future. The journey along the freeway, like the earlier journey of the frontiersman along the river, alluding to romantic implications of freedom and opportunity, has been transformed by Maria. She uses the road as an escape. Barbara Harrison comments:

> Remembering is to express a wish to be dead, to return to some pre-Edenic state in which good and evil, right and wrong, do not exist. It is a wish to erase, not only one's personal painful past but our collective past – which, in turn, is an invitation to believe that we cannot, individually or collectively, affect the present or the future.[62]

Maria erases all thoughts about living in the future, and to remember is to acknowledge that she has a past, something she is determined not to do. On the freeway, the people, the places and the events of her life can be forgotten, if only for a short period of time.

As wife of a Hollywood film executive, Maria's life is picture-perfect, the superficial surface never revealing the painful truth of what lies beneath the external layering. As Maria sits poolside, talking to her friend BZ, separated from her husband and with her daughter in a care facility with mental illness, she imagines:

> It was the hour when in all the houses all around the pretty women were putting on perfume and enamelled bracelets and kissing the pretty children goodnight," and "the water in the pool was 85. The water in the pool was always 85 and the water was always clean.[63]

Hollywood is filled with images of people living out perfect existences in perfect houses but reality is far from this silver-screen induced idyllic fantasy. Maria constantly uses media-created visions of domestic bliss to imagine a future filled with familial happiness and to further forget her past. She dreams, "every morning in that house she would cook while Kate did her lessons." In these dreams, "there were only three people and none of them had histories."[64] In Hollywood and in Los Angeles, the past is irrelevant and Maria knows it. Image is everything and as long as you are willing to make-believe, anything is possible. Didion in *Slouching Towards Bethlehem* comments:

> The future always looks good in the golden land, because no-one remembers the past. Here is where the hot wind blows and the old ways do not seem relevant, where the divorce rate is double the national average and where one person in every thirty-eight lives in a trailer. Here is the last stop for all those who come from somewhere else, and all those who drifted away from the cold and the past and the old ways. Here is where they are trying to find a new life style, trying to find it in the only places they know to look: the movies and the newspapers.[65]

If the reality of a depressing past is too difficult to cope with, the imagined future of the Los Angeles lifestyle means that the past can be simply swept away, replaced or forgotten.

Maria tries to forget her past, tries to imagine a brighter future but she cannot escape the constant presence of her despair at the meaninglessness of life.

The loss of her mother in a car accident, the loss of her daughter Kate to mental illness and the loss of her husband Carter due to her own mental instability have destroyed any hope Maria holds for the future. Without the comfort of a family unit, Maria is constantly searching for the security of a home life that does not exist outside her imagination. She replays the version of her mother's last few moments, one she has created for herself, over and over, hoping her thoughts were with her, knowing with almost certainty they were not. Maria's love for Kate is desperate, attempting to connect to her daughter in a way she never could with her own mother. She ignores the advice of the doctors and Carter to give Kate time to adjust to her illness, instead relying on the impossible dream of domestic stability she desperately wants to provide. Carter remains a presence in Maria's life but apart from his inclusion in the deluded images of a happy family future, their marriage is over long before the legal process of divorce begins. Carter is different from Maria in that he does not share her nihilistic views and finds meaning in his work with movies. He cannot relate to, or understand what Maria is experiencing. When she asks him, "why do you say those things? Why do you fight?" he answers, "to find out if you are alive."[66] To fight is to show signs of life but Maria refuses to do even that.

While the freeway is a vehicle for Maria to forget the past, the overpowering influence of the desert in *Play It As It Lays* is impossible to avoid. Like the ocean and mountain ranges, the desert surrounds Los Angeles, making the city a difficult place to get to, but making it a harder place to leave. It is also Maria's home. Charles Crow comments:

> The desert, lurking in the east, is the city's doom. Out of the desert blow the Santa Ana
> winds, drying the hillsides and spreading the brush fires which smoulder in the background

> of the novel...Deserts always have been places of prophecy and truth-seeking, and the message of this desert, "the hard empty white core of the world," is annihilation, nothingness.[67]

Growing up in the comically tiny Silver Wells, Maria's gambling father tried to turn the town into a destination for travellers but failed. The town disappears, replaced by a missile testing site, further evidence for what Maria sees as a bleak and desolate future, a precursor to the atomic future feared by the Beats and Marion Faye. Maria does not have any dislike for her birthplace. Silver Wells is simply another example of the emptiness and nothingness in life. Chip Rhodes comments:

> It is not that she hates where she comes from, such hatred the standard motivation of the bildungsroman protagonist who leaves a stifling small town for the big city. Rather, she cannot have any strong feelings about Silver Wells because it has no human community and no standards and regulations against which to rebel.[68]

Her family and her hometown have died or disappeared, the memories she has only reinforcing her belief in the meaningless nature of her existence. BZ shares Maria's desperate vision of the future. When Maria asks him, "don't you ever get tired of doing favours for people?" BZ replies, "you don't know how tired."[69] Disillusioned by the pretext of success that defines an individual in Hollywood and in Los Angeles, the ability to find meaning in life is no longer possible. Instead, escape only comes in the form of death; those left behind will choose to accept their position or continue to act out a life that is more illusion than reality.

Whether the individual is caught up in a series of inexplicable events that only create more questions, or chooses to ignore the reality of life as it exists around them, Los Angeles continued to isolate its inhabitants. The landscape of the city, while attempting to modernise, replicate and improve, achieves only fragmentation and disillusionment through its transparency and contradiction. People become lost in the city, trying to find a constant in the fast-paced, ever-changing world around them. History is manipulated to encourage

growth and discourage unrest. As cultural and social divides widen to create a new friction

and tension within Los Angeles, the literature intensifies the alienation, loss and

occasionally the playful aspects of a post-modern world.

[1] Edward Soja, *Postmetropolis: Critical Studies of Cities and Regions* (Malden, MA: Blackwell Publishers, 2000), p. 132.

[2] Charles Higham and Joel Greenberg, *Hollywood in the Forties* (New York, NY: A. S. Barnes and Co. Ltd, 1968), p. 14.

[3] Evelyn Waugh, *The Loved One* (Bristol, UK: Chapman and Hall, 1958), p. 32.

[4] Ibid, p. 33.

[5] Ibid, p. 35.

[6] Walter Wells, 'Between Two Worlds' in *Los Angeles in Fiction* (Albuquerque, NM: University of New Mexico Press 1985), pp. 187-206.

[7] Evelyn Waugh, *The Loved One*, p. 36.

[8] Ibid, p. 1.

[9] Ibid, p. 3.

[10] Ibid, p. 28.

[11] Ibid, p. 75.

[12] Ibid, p. 118.

[13] Ibid, p. 96.

[14] Ibid, p. 117.

[15] Nathanael West, *The Day of the Locust*, p. 42.

[16] Fredrick L. Beaty, *The Ironic World of Evelyn Waugh – A Study of Eight Novels* (DeKalb, Illinois: Northern Illinois University Press 1992), p. 173.

[17] Malcolm Bradbury, 'America and the Comic Vision' in *Evelyn Waugh and His World* ed. by David Pryce-Jones (London, UK: Weidenfeld and Nicolson 1973), p. 168.

[18] David Fine, *Imagining Los Angeles*, p. 173.

[19] Norman Mailer, *The Deer Park* (New York, NY: Howard Fertig 1980), p. 4.

[20] Ibid, p. 4.

[21] Ibid, p. 47

[22] Ibid, p. 124.

[23] Ibid, p. 171.

[24] Ibid, p. 191.

[25] Mark Winchell, 'Fantasy Seen: Hollywood Fiction Since West' in *Los Angeles in Fiction*, pp. 165-86.

[26] Norman Mailer, *The Deer Park*, p. 136.

[27] Ibid, p. 205.

[28] Mark Winchell, 'Fantasy Seen: Hollywood Fiction Since West' in *Los Angeles in Fiction*, pp. 165-86.

[29] Mailer, p. 161.

[30] Allen Ginsberg, 'Howl' in *'Howl' and Other Poems* (San Francisco, CA: City Lights Books, 2000), p. 17.

[31] John Tytell, *Naked Angels* (New York, NY: McGraw-Hill Book Company, 1976), p. 4.

[32] J. Michael Lennon, 'A Conversation with Norman Mailer', *New England Review* (Middlebury, NE: 1999), p. 139, in *ProQuest* <http://proquest.umi.com/pqdweb?did=43693347&sid=2&Fmt=3&clientld=8119&RQT=309&VName=PQD > [accessed 20 October 2006].

[33] Norman Mailer, 'The White Negro', in *Advertisements For Myself* (London, UK: Andre Deutsch Ltd., 1959), p. 283.

[34] Ibid, p. 284.

[35] Ibid, p. 296.

[36] Allen Ginsberg, 'Howl', p. 11.

[37] Jonah Raskin, *American Scream* (Los Angeles, CA: University of California Press, 2000), p. xxi.

[38] Gregory Stephenson, *The Daybreak Boys: Essays on the Literature of the Beat Generation* (Carbondale, Ill: Southern Illinois Press, 1990), p. 53.

[39] Allen Ginsberg, 'Howl', p. 22.

[40] Gregory Stephenson, *The Daybreak Boys: Essays on the Literature of the Beat Generation*, p. 57.

[41] Allen Ginsberg, 'Howl', p. 28.

[42] Edward Mendelson, *Individual and Community: Variations on a Theme in American Fiction* (Durham, NC: Duke University Press, 1975), p. 186.

[43] Geoffrey Lord, 'Mystery and History, Discovery and Recovery' in Thomas Pynchon's *The Crying of Lot 49* and Graham Swift's *Waterland*, *Neophilologus*, 81 (1997), p. 145.

[44] Terry Eagleton, 'The Illusions of Postmodernism' in *University of Louisville – Fulbright Summer Institute on Contemporary American Literature* (Oxford, UK: Blackwell, 1996), pp. 27-44.

[45] Thomas Pynchon, *The Crying of Lot 49* (New York, NY: First Perennial Classics, 1999), p. 14.

[46] Ibid, p. 14.

[47] Patricia Burgh, '(De)constructing the Image: Thomas Pynchon's Postmodern Woman' in *Journal of Popular Culture*, 30 (1997), p. 13.

[48] Thomas Pynchon, *The Crying of Lot 49*, p. 28.

[49] Ibid, p. 29.

[50] Ibid, p. 30.

[51] Geoffrey Lord, 'Mystery and History, Discovery and Recovery' in Thomas Pynchon's *The Crying of Lot 49* and Graham Swift's *Waterland*, p. 153.

[52] Tony Tanner, *Thomas Pynchon* (New York, NY: Methuen and Co. Ltd 1982), p. 65.

[53] Thomas Pynchon, *The Crying of Lot 49*, p. 36.

[54] Ibid, p. 37.

[55] Ibid, p. 126.

[56] Ibid, p. 118.

[57] Ibid, p. 41.

[58] Ibid, p. 101.

[59] Richard Lehan, 'The Los Angeles Novel and the Idea of the West' in *Los Angeles in Fiction*, pp. 29-41

[60] Joan Didion, *Play It As It Lays* (New York, NY: Penguin Books Ltd 1973), p. 18.

[61] Joan Didion, 'Bureaucrats' in *The White Album* (New York, NY: Simon and Schuster 1979), pp. 79-85.

[62] Barbara Harrison, 'Joan Didion: The Courage of Her Afflictions', *The Nation*, 9 (1979), p. 278, in *Literature Resource Center*
<http://galenet.galegroup.com.ezproxy.waikato.ac.nz:2048/servlet/LitRC?vrsn=3&OP=contains&locID=waikato&srchtp=athr&ca=1&c=2&ste=16&stab=512&tab=2&tbst=arp&ai=U13026265&n=10&docNum=H1100002218&ST=didion%2Ctjoan&bConts=16303> [accessed 1 August 2006] (para.3 of 23).

[63] Joan Didion, *Play It As It Lays* p23

[64] Ibid, p. 91.

[65] Joan Didion, 'Some Dreamers of the Golden Dream' in *Slouching Towards Bethlehem* (New York, NY: Farrar, Straus and Giroux 1990), pp. 3-28.

[66] Didion, *Play It As It Lays*, p. 49.

[67] Charles Crow, 'Home and Transcendence in Los Angeles Fiction' in *Los Angeles in Fiction*, pp207-23

[68] Chip Rhodes, 'The Hollywood Novel: Gender and Lacanian Tragedy in Joan Didion's *Play It As It Lays*', *Style*, 34.1 (2000), in Academic Premier Search
<http://search.ebscohost.com.ezproxy.waikato.ac.nz:2048/login.aspx?direct=true&db=aph&AN=3667289&site=ehost-live> [accessed 16 February 2007] (para.10 of 39).

[69] Didion, p. 34.

CHAPTER THREE

```
FADE IN:

OVER the opening strains of "I LOVE YOU, CALIFORNIA," a

MONTAGE: a mixture of headlines, newsreel footage and live

action.  Economy Booming!  Post-war Optimism!  L.A: City of

the future!  But most prominent among them: GANGLAND!  Police

photographers document crime scene.
```
Brian Helgeland – *L.A Confidential* (1997)

From the outset of the California booster's promotion of Los Angeles as a city

bathed in sunshine and situated on the Pacific Ocean, the outsider's image has been tinged

with a glorified vision of the Hollywood Hills, of the Westside or of the coastline, replete

with bright, happy, opportunity-laden locations. Yet the vast majority of Los Angeles lies

inland often overlooked and repeatedly airbrushed out by the media controlled pictures of

the 'Golden Land.' Reyner Banham, in *Los Angeles - The Architecture of Four Ecologies*

refers to the inland areas of Los Angeles as "The Plains of Id," an, "endless plain endlessly

gridded with endless streets, peppered endlessly with 'ticky-tacky' houses clustered in

indistinguishable neighbourhoods, slashed across by endless freeways…"[1] These areas of the city, the South Central and East Los Angeles neighbourhoods are often hidden away to

ensure the continued front of a prosperous and safe environment. This fantasy betrays the

reality of the city, hiding the tension and diversity of the sprawling metropolis through the

continuing conflict between the real and the imagined. Edward Soja comments:

> On location since the 1920's, is the multitude of "dream factories" that compromise what is
>
> still called "the Industry," mass-producing moving pictures of Los Angeles that insistently
>
> substitute reel stories for real histories and geographies. Camera crews "shooting" scenes
>
> depicting practically every place on earth (and often off-earth) are a familiar sight on the
>
> streets of the city, and a constant local reminder of the confusing interplay between fantasy
>
> and reality that pervades everyday life in the City of Angels.[2]

Opposing the fantasy of Los Angeles, where, "in 1960, more than 80 percent of the

inhabitants were non-Hispanic whites,"[3] is the reality of a multi-racial city. Minorities have

historically been misrepresented, misunderstood and many times, mistreated. From the

confinement of Japanese-Americans in 1942 under Executive Order 9066[4] (German-

American and Italian-Americans were conveniently overlooked), through the Sleepy

Lagoon murder trials of 1942, Zoot-suit riots of 1943[5], the 1965 Watts riots to the Rodney

King initiated riots of 1992[6], minority groups have been isolated and recognized only by

the sporadic, violent episodes splashed across newspapers and television screens. As media

coverage exposed the real Los Angeles to wider national and international audiences,

earlier attempts to 'whitewash' the city began to unravel, and writers turn their attention to

forgotten parts of the city.

The disconnected nature of Los Angeles and the influence of its media sources in

print, film and television change, manipulate and confuse a single event until fact blurs into

fiction and one person's truth becomes another's lie. The Rodney King trial and the O.J

Simpson murder trial of recent times, both with strong racial undertones, were two events

saturated by the media, who used long lists of witnesses, the testimony of 'experts' and the stories of victims until the viewers were unable to distinguish between fact and fiction, real

and imagined news. While Los Angeles, with the aid of Hollywood, still remains a city

with a promising future, alongside the economic booms and optimism that have highlighted

the city's history is a shared realization of an equal amount of violence, prejudice,

depression and shattered dreams. Using the past, present and the future to reveal a

perspective in stark opposition to the imagined Los Angeles, the literature reflects the

changing face of the city. Acknowledging the massive suburban shift as the city continues

its rapid growth begun in the 1920's, writers dealing with issues of racial and economic

disparity and moral corruption that haunted the city were given prominence.

In the hard-boiled vein of Los Angeles detective story-telling, James Ellroy's *The

Black Dahlia* ignores sanitized pictures of the city and sunshine and smiles, choosing to

once again explore the mean streets corrupted and polluted by criminals and city officials

alike. Framing the action around two historical events, the Elizabeth Short murder of 1947

and the Zoot-suit riots of 1943, *The Black Dahlia* blurs the lines between fact and fiction.

Using true events and fictional characters, Ellroy writes about the past in order to

understand the confusion of the present. Influenced by the loss of his own mother to an

unsolved murder, *The Black Dahlia* becomes both "a salutary ode to Elizabeth Short and a

self-serving and perfunctory embrace,"[7] of his mother, and therefore, his own identity.

Elizabeth Short's mutilated body was found on the morning of January 15, 1947 in

central Los Angeles:

> It was the body of a woman who'd been cut in half and was laying face-up in the dirt. The
>
> woman's arms were raised over her head at 45-degree angles. Her lower of half was
>
> positioned a foot over from her torso, the straight legs spread wide open. The body
>
> appeared to have been washed clean of blood, and the intestines were tucked neatly under
>
> the buttocks.[8]

When FBI examiners matched the fingerprints to a beautiful, 22-year-old aspiring actress, the media responded with sensationalized headlines and reports, seizing on the story of an East coast girl murdered while trying to make it in Hollywood. Short's legend grew as the media fed off her tragic death, the factual evidence of her life became unimportant as an imagined life grew in stature. Larry Harmisch comments:

> The facts became something that people wanted to believe, instead of what happened. It became a myth, a cautionary tale about Los Angeles, how it chews up and spits out star-struck girls.[9]

Ellroy uses the ambiguity behind the factual evidence collected from the real Elizabeth Short's murder to present his case of corruption and mistrust in *The Black Dahlia*.

The Black Dahlia begins with Dwight 'Bucky' Bleichert, a junior police officer, who is called to duty in the midst of the Zoot-suit riots.

> Word hit the papers that the zooters were packing Nazi regalia along with their switchblades, and hundreds of in-uniform soldiers, sailors and marines descended on downtown Los Angeles, armed with two-by-fours and baseball bats...Every central Division patrolman was called in to duty, then issued a World War I tin hat and an oversize billy club known as a nigger knocker.[10]

Between the L.A.P.D, the media, the attackers and the victim's accounts, what is considered the truth becomes difficult to separate from fiction. The war moves into Los Angeles on the suspicion of Nazi sympathizers infiltrating the 'zooters' ranks. With the highly-charged atmosphere of suspicion and prejudice that had been generated by America's entry into World War II, rumours of 'zooters' carrying Nazi regalia only fueled the racial fire and promoted the use of arbitrary violence condoned by a patriotically-projected police department.

The zoot-suit had become a popular clothing statement in the 1940s, and a means by which black and Mexican-American youth expressed their individual and creative rights.

With the bombing of Pearl Harbour and Los Angeles' proximity to the Pacific Ocean, the

Zoot-suit riots became a racially motivated and an excuse to use excessive force. Douglas

Daniels comments:

> The very terms in which the Los Angeles riot was discussed were totally misleading,
>
> indicating the desire of the police and the press to recast events to fit their biased views of
>
> society. The zoot-suit rioter was neither Mexican-American, Mexican nor black delinquent
>
> youth; the actual rioters were in fact the white U.S soldiers and sailors from nearby military
>
> bases and white Los Angeles residents who hunted down zoot-suiters and beat and stripped
>
> them of their clothing.[11]

Yet to the outside world, the media cast the 'Zooters' as black or Mexican villains, laying

blame on them for the destruction of public property and disruption of civic order in the

city. The immersion of Bleichert into this historical event immediately shows him as

unable to control the situation, as the damage is too widespread. While physically strong,

Bleichert is exposed as vulnerable to the world around him. Ellroy comments:

> He's standing outside momentous events. He's lost in scrutiny. He wants to control. He
>
> wants to capitulate. His inner-life is near chaotic. He needs to impose external order to
>
> countermand his mental state.[12]

Confronted by a very real situation created from an entanglement of misinformation and

hatred, Bleichert is, "terrified because the good guys were really the bad guys."[13] Trying to

regain order is impossible. Sent in to protect the city from an expected attack from Nazi-

loving zoot-suiters, Bleichert instead watches, "sailors kicking in windows all along

Evergreen," and marines, "systematically smashing streetlights, giving themselves more

and more darkness to work in." Furthermore, Bucky watches his, "fellow officers

hobnobbing with Shore Patrol goons and MPs,"[14] choosing to ignore the blatant

contradiction between the information fed to the public by the media and police, and the

actual events taking place. Ellroy's Los Angeles is never simply black or white and his characters are never completely good or bad. Every man and woman has a history, or a past with a dark secret.

The city, through Deputy D.A. Ellis Loew decides to create the working partnership of Bleichert and Officer Lee Blanchard, both prize fighting boxers, manipulating public opinion through the use of the media. Recognizing the attraction of Bleichert and Blanchard as local sporting heroes, Loew organizes a one-off fight against each other, generating Department interest and capturing the imagination of the public. As Loew admits, "Publicity all the way, straight to the '47 Special Election."[15] The two boxers become known as 'Fire and Ice,' recognizable to the average citizen and creating a more accessible relationship between the Department and the people of Los Angeles. The fight becomes a spectacle; all parties get what they deserve and the result appears to benefit everyone. Bleichert however, has sold out Loew by deliberately losing the fight in a late round, choosing to take a dive to collect on a bet against winning. It costs him the chance at promotion promised by Loew for winning the fight, but releases some of the guilt he has over a sick German father and for the forced informing on Japanese-American friends in his earlier days.

The murdered body of Elizabeth Short and the attempt to solve her case is also seized upon by the Department and DA's office as a further chance for political advancement and public support within the novel. Bleichert recognizes it instantly commenting, "It's a showcase. A nice white girl gets snuffed; the Department goes all out to get the killer to show the voters that passing the bond issue got them a bulldog police force."[16] With the media clambering to sensationalize the murder as a gruesome attack on an innocent white girl with headlines like 'Hunt Werewolf's Den in Torture Slaying!!!,'[17] Short becomes a political pawn used to improve the Police Department's image of upholding law and order in the Los Angeles community.

As Bleichert becomes more involved with the case at the insistence of his partner, conflicting evidence, suppressed information and wrongful arrests uncover overwhelming corruption within the police force. The 'Black Dahlia' murder becomes a small part in a much wider maze of criminal activity involving high-ranking city officials and influential businessmen, from extortion and racially motivated violence to sexual episodes with prostitutes, one of whom is identified as Elizabeth Short. The L.A.P.D as a symbol of civility, order, truth and justice is exposed as fraudulent and Ellroy's Los Angeles shown to be a city run by criminals, the media churning out falsified stories to cover their tracks. Like Sergeant Jack Vincennes in *L.A Confidential* selling information to 'Hush Hush' magazine and advising stage actors how to behave like 'real' police officers, the truth is often distorted; insider information can be bought for a price and reality hidden from the public.

Yet, for Bleichert the 'Black Dahlia' murder is a personal conflict of unstable relationships and an attempt to forget his own past. Blanchard, overcome by the memory of his own sister disappearing, obsesses over catching the murderer and falls apart emotionally. He goes to Tijuana to escape the madness of the city and Bleichert believes he does it rather than accept the duplicity of their superiors. While in Tijuana himself investigating a pornography ring involving the 'Black Dahlia' Bleichert uncovers Blanchard's own demise and his secret past as accessory to a string of bank robberies. Blanchard is as corrupt and guilty as the criminals he has chased. Like Oedipa before him, the more Bleichert discovers, the less he understands. As *The Black Dahlia* unravels, the mystery deepens and more seemingly innocent people are implicated in illegal activity. Bleichert must face the fact he was, "best friend to a bank robber,"[18] and his lover Kay pleads with him to prevent the truth from coming out because, "we can't drag his name down. We have to give it all up and forgive him and get on."[19] The overwhelming confusion of the situation forces him to leave her in order to try and make sense of his own life.

In the ongoing façade of a white L.A.P.D intent on maintaining an appearance of respectability, Bleichert slowly separates fact from fiction as he comes closer to identifying the killer of Elizabeth Short. But, as clues continue to implicate the people he is intimate with, the reality of exposing his own indiscretions with the capture of the guilty party becomes apparent. Carole Allamand comments:

> Ellroy's detective is always a double agent: tracking down "criminals" he is merely covering his own tracks, which lead the reader back to the original murderer, the storyteller himself, and the false evidence that is his work.[20]

Bleichert's own secret sexual relationship with the daughter of a powerful real estate speculator leads to the cover-up of a potentially important witness. The trail of the murderer back to her family make Bleichert as guilty as the criminal and he acknowledges his own fallibility. There are no innocent parties in Ellroy's Los Angeles. From wealthy property owners to high-ranking city officials, from Police Plaza to Skid Row, corruption is all-encompassing. The media is manipulated and paid to present a very different picture of the city, and no-one is beyond the scope of greed and deception.

In similar vein, Joseph Wambaugh's *The Onion Field* examines the true story of two Los Angeles police officers who are kidnapped by two criminals during a routine car stop in 1963. *The Onion Field* is told from the perspective of not only the police officers but also from that of the criminals. The upbringing and childhood influences of the four men involved, the shooting of police officer Ian Campbell, and the following court case present an intricate collection of historical fact, media speculation and attributing social factors. Los Angeles is again seen as a disjointed, separated city and the people and places, while sounding like something pulled from a Hollywood script, are very real, but, like

Ellroy, Wambaugh attempts to unravel the past as a means of understanding the present, despite a media that corrupts all meaning.

The picture of Hollywood on March 9, 1963 is far removed from the sunshine-kissed image of Southern California normally associated with the capital of the film world. This Hollywood, with, "intersections where traffic snarled," and, "streets undulated with out-of-towners, roaming groups of juveniles, fruit hustlers, desperate homosexuals, conmen, sailors, marines,"[21] is home to more than movie stars. "Nothing the Hollywood Chamber of Commerce said could camouflage the very obvious dangers to tourists on those teeming streets,"[22] and Hollywood is suddenly revealed as a dangerous and unpredictable place, of the kind normally associated with the downtown and central city areas. It is in this environment that Gregory Powell and Jimmy Smith, both life-long hustlers and criminals, kidnap Campbell and Karl Hettinger. The subsequent shooting of Campbell and the attempted shooting of Hettinger shock the WASPish (White Anglo-Saxon Protestant) L.A.P.D who pride themselves on the principle of upholding the laws of the city. To the police, "there had always been rules in the game. One had to have a good reason to kill a cop…In killing Ian Campbell they had scoffed at fair play, scorned the rules of the game."[23] By killing a police officer, without reason, the illusion of invincibility within the department is shattered. With the media coverage following what would become the longest criminal case in California history; the public was made aware that even the supposed tourist destinations of Los Angeles, like Hollywood, were no longer safe.

In exploring the background story of the two criminals, Smith and Powell, the reader is introduced to a different, darker side of Los Angeles. Jimmy Smith's neighbourhood is one where, "the winos roamed west towards downtown, towards the live ones, and left Fifth and Stanford to the blacks, the pimps and prostitutes, the bootleggers, dope dealers, the thieves and con men."[24] This is the polar opposite of a Beverly Hills lifestyle, where, "the boy came twice a week to vacuum the pool and the man came four times a week to work on the roses."[25] The forgotten, hidden parts of Los Angeles are home to desperate people whose dreams have not come true. An uneducated black man with a criminal record like Jimmy Smith is feared by the people from the burgeoning suburbs and affluent areas of Los Angeles who would rather ignore and forget his existence and the neighbourhood he comes from. An associate of Smith's comments, "You livin in Los Angeles man. They's afraid of you. They thinks you is some wild nigger. Stare at em and watch."[26] Like the killers, Dick and Perry, in Truman Capote's *In Cold Blood*, Smith and Powell are humanized by the author's investigation into their lives. Their own weaknesses and fears are examined and the harsh reality of the murder they commit becomes more terrifying to the reader through a proximity to their inner minds.

The trial following the arrests, and the public fascination with the case, distort the lines between fact and fiction, tragedy and farce. A fellow inmate of Powell's who writes a drama about the night of the shooting becomes part of the defence team's evidence to show Powell as a victim of society, whose, "wishes were useless for I knew I had been born to lose."[27] Manipulating the legal system, the defence team attempt to prolong the proceedings, objecting every point and continuously calling for a mistrial. It becomes so tedious that the original prosecutor complains, "There's no concept of what's gone before. No one cares that a cop was killed so brutally."[28] The spectacle means the crime is now secondary to the antics performed by the defence team. Forty years later in a California courtroom, lawyers for O.J. Simpson would divert attention away from the crime of a double-murder to focus on the alleged racial corruption of the investigative police team and an ill-fitting glove. A trial that should have seen both Smith and Powell executed for their crime, becomes, like Ian Campbell's funeral, "a goddamned three-ring circus,"[29] where the public are more interested in the show than the tragic reality of the events.

Inland from the beaches of Malibu and the mansions of the Hollywood Hills are the gritty, mean streets of Walter Mosley's hard-boiled detective novels. And while Mosley maintains the tradition of the hard-boiled detective established by Hammett and Chandler, his protagonist and location have a distinct difference. Ezekiel 'Easy' Rawlins is a black man and his adopted community is in South Central Los Angeles, an area of the city scratched upon but often bypassed in earlier detective novels. Roger Bergen comments:

> For Mosley, South Central L.A is not merely an exotic location – or, worse, a plot device to begin a novel. Rather, it is the community where Mosley's novels are set. In a sense, Mosley elevates black L.A in his novels into a significant location.[30]

In opening up another part of Los Angeles unknown to many people outside of the community, Mosley's writing highlights the diversity of a city with a growing reputation for being fragmented and divisive. It also presents Rawlins as an African American hero in a genre and a city previously dominated by white detectives like Chandler's Philip Marlowe. Helen Lock comments:

> The "changing implications" of the investigative process become infinitely more complex, and painful to negotiate, when a black detective finds himself haunting an additional borderland, that where the interests of his own community and those of the broader, predominately white, society uneasily co-exist and frequently collide.[31]

While Rawlins may be an uneasy hero, importantly he represents a black man upholding the hard-boiled code of justice, a code that had not been examined in detail within South Central Los Angeles.

Easy Rawlins is an educated black man trying to maintain a balance between keeping in touch with his community and distancing himself from the criminal element of his work. Rawlins is also aware of the dangers associated with being a black man in Los Angeles. He distrusts the police and is conscious of the prejudices people automatically place on him. In *A Little Yellow Dog*, Easy works as a supervising custodian at a Junior High School, prepared to leave the streets behind as a memory of his past life. His new dream, like many others, is to raise his children and to own a home. The difference is this is not a typical suburban Los Angeles dream. Rawlins' two children are adopted orphans from abusive backgrounds and the location of his house is in a Watts neighbourhood that only eighteen months later would be a central point in one of the most explosive civil riots in American history. To many residents of Los Angeles, South Central and Watts are alien, territory associated with poverty, gang violence and drugs. David Fine comments, "Watts is terra incognita, a city apart, a black city within a white city, unknown and incomprehensible to them."[32] Mosley's novels allow the reader to discover this forgotten, "incomprehensible" part of the city, his stories reminding us that corruption and crime reach beyond the physical boundaries and colour of a person's skin.

The feeling of distrust Rawlins has for white people in a position of authority is a combination of his experience on the street and a remembered history of mistreatment that goes back generations. When the principal of the high school brings Easy into his office to accuse him of stealing, Rawlins says, "Newgate was watching me. I was used to it. White people like to keep their eyes peeled on blacks, and vice versa. We lie to each other so much that often the only hope is to see some look or gesture that betrays the truth."[33] There is a general understanding that any criminal activity will automatically arouse suspicion toward Rawlins and he is careful not to show any sign of weakness, even when he has done nothing wrong. The same kind of caution is shown in Rawlins's driving habits around the city. "I took a southeastern route because that was the 1960s and black men couldn't take a leisurely drive in white Los Angeles without having the cops wanting to know what was going on."[34] Anything that appears suspicious, particularly involving a black man in a predominately white area of Los Angeles, is asking for trouble. Easy Rawlins, an innocent man, chooses to avoid confrontation, preferring to remain invisible whenever possible.

Jimmy Smith and Greg Powell in *The Onion Field*, guilty of previous crimes, do not take

the educated outlook of Rawlins to remain incognito and looking, "as though they'd driven

off the Columbia Pictures lot farther south on Gower,"[35] become an obvious target for

police intervention. Maintaining a low profile and defensive attitude allow Rawlins to

move between the black and white communities with as little friction as possible.

Location is an important part in *A Little Yellow Dog*, depicting Watts as a very

different part of Los Angeles but still showing the problems of the people as relative to the

city as a whole. The "definition of a loser in L.A: a man without a car,"[36] is the same in

Watts as in Santa Monica. Yet the description of Rawlins's part of the city with, "palm

trees and poverty; neat little lawns tended by the descendants of ex-slaves and massacred

Indians,"[37] drives home the racial divide that separates the city. It is often easier for these

parts of the city to be ignored than to be brought out into the open.

The media chooses not to report on the violence on black people. In *A Little Yellow

Dog*, the death of two black American brothers in Los Angeles receives no mention in the

newspaper but the death of several hundred Haitians in a flood thousands of miles away

makes front page headlines. "To the white press, and many white Americans, black people

were easier to see as exotic foreigners, somebody else's people. But the lives of black

Americans were treated with silence."[38] There is little interest in reporting the death of a

black man but an act of violence by a black man or the suspicion of a crime committed

attracts a great deal of attention. When Rawlins is asked to accompany a police officer to

the Hollywood station, he witnesses, "thirty or so men living in cages underground like

livestock waiting for some further shame to be laid on them. Like sharecroppers or slaves

living in shanty shacks on the edge of a plantation."[39] The crimes these men have

committed, or are accused of committing do not make headlines but they are still treated in the harshest manner. Rawlins is shown these cages as a warning of what could happen to him if he refuses to co-operate.

By the time the Hughes brothers' film, *Menace II Society*, was released in 1993, Los Angeles had experienced the 1965 Watts riots and the 1992 rioting following the sentencing of the police officers charged with beating Rodney King. Los Angeles in *Menace II Society* is a community dominated by violence, guns and the financial rewards of selling drugs. The images are a shocking contrast to the typical Hollywood portrayal of Los Angeles with calming ocean breezes and swaying palm trees. Paula Massood comments:

> It is through the process of inclusion and exclusion that Hollywood has helped L.A to nurture and reify a particular set of urban signs – palm trees, sun, abundance, paradise. Yet the manufacture of one particular group of images has as its mirror the exclusion of areas which do not meet the criteria for this imagined city.[40]

These areas are populated not only by criminals but by the victims of a media-driven society eager to ignore the tenuous living conditions of their close neighbours in South Central. As is the case with Walter Mosley's Los Angeles, the spaces in *Menace II Society* do not nurture the Hollywood image but present an important portrayal in showing their part of the city as crowded, dangerous and embarrassingly poor.

The reversal of an earlier dream of Southern California, of the dreamer coming to Los Angeles to find success and fortune is adopted in the 'hoods of South Central. The two ways to succeed are either submitting to a life of crime and violence, or to escape. South Central is viewed as a modern wasteland; the lucky ones make plans to go to Atlanta for work, or to Kansas to play football. The unlucky ones turn to gangs and drugs. The residents of the 'hood in *Menace II Society* are, according to one gang member, "America's nightmare – young, black and don't give a fuck."[41] Even a respected school teacher is resigned to being constantly harassed and downtrodden saying, "the white man is always on your back but what are you going to do.'"[42] This honest admission solidifies the very

different city of Los Angeles in *Menace II Society*. Massood comments:

> It is a legacy in which the city has been mythologized as both a utopia – as a space
> promising freedom and economic mobility – and a dystopia – the ghetto's economic
> impoverishment and segregation.[43]

Some people are still living the California dream and enjoying the opportunity for

economic success but the people of Watts and South Central L.A struggle everyday against

poverty and prejudice.

The opening up of the Los Angeles landscape, through the literature of writers like

Walter Mosley, and the increased awareness raised by the civil rights movement, coincided

with the strengthening of the Chicano movement in the 1960s and 1970s. While Mosley's

books and films like *Menace II Society* brought an insider's perspective to the streets of

South Central, works like Oscar Zeta Acosta's *The Revolt of the Cockroach People*

journeyed into the *barrio*, revealing East Los Angeles and a concealed Mexican culture. As

with the South Central area of the city, the *barrios* of East Los Angeles were seldom

pictured to represent the qualities of Southern California people had come to expect.

Raymond Paredes comments:

> Los Angeles has been, since the mid-nineteenth century, a city uneasy about its Mexican-
> ness. The first several generations of Anglo settlers sought either to obliterate the
> prevailing Mexican character of the city or to make it over in their own preferred
> conceptualization, both responses deriving from the widespread belief that actual Mexican
> culture was unworthy of recognition and preservation.[44]

It was this attitude of destroying the past that stigmatized the Mexican American culture of

Los Angeles and prevented the people from publicly and equally expressing their culture

and rights.

By the time of the first surge of growth during the 1920s, Mexican culture had become either submerged or reduced to specific neighbourhoods. The romantic version of the region's history was much easier to swallow. Ruben Martinez comments:

> Such is L.A's acute case of amnesia. Better to think that Spanish troubadours on palomino horses swept castanet-snapping *senoritas* off their feet rather than accept the fact that the *Californios* were actually *mestizo* Iberophiles who exploited the region's majority population of poor Mexicanos and smattering of Native Americans.[45]

The Sleepy Lagoon murder case in 1942 and the Zoot-suit riots in 1943 were two examples of the WASP dominated media and law enforcement officials influencing public opinion and falsely laying blame on people of Mexican descent. Mexican culture in Los Angeles during this period received no public acknowledgement, the media only showing interest in events that influenced the surrounding areas of the barrio. Octavio Paz, visiting Los Angeles from Mexico in the 1940s commented:

> This Mexicanism – delight in decorations, carelessness and pomp, negligence, passion and reserve – floats in the air. I say "floats" because it never mixes or unites with the other world, the North American world based on precision and efficiency…It floats, never quite existing, never quite vanishing.[46]

The Booster past of the city affected the Mexican people, altering their perception and confidence. Paz comments, "They have lived in the city for many years wearing the same clothes and speaking the same language as the other inhabitants, and they feel ashamed of their origin; yet no-one would mistake them for authentic North Americans."[47] The loss of pride and identity with a cultural past reflected on a larger scale a city eager to forget its checkered past and one trying to establish a positive, economically prosperous future.

When Acosta's Buffalo Brown, in *Revolt of the Cockroach People*, arrives in Los Angeles in January 1968, there is the beginnings of the Chicano movement, which alongside the civil rights movement, was a defiant stand to establish equality for the fast-growing Mexican population of the city. Acosta's intentions are at first simple and selfish, "to find "THE STORY" and write "THE BOOK" so that I could split to the lands of peace and quiet where people played volleyball, sucked smoke and chased after cool blondes."[48] Like so many others, Brown has been manipulated by the imagined version of Los Angeles as a peaceful, carefree city. After six hours in Los Angeles he already feels duped, "my bones have told me that I have come to the most detestable city on earth. They have carried me through the filthy air of a broken city filled with battered losers."[49] Fooled into coming to a city appearing to make dreams come true, the downtown area seems a lifetime away from the sun and surf images the outside world equate to Los Angeles. There are no movie stars or beautiful blondes, only, "streets filled with dark people, hunchbacked hobos, bums out of work, garbage of yesterday and tomorrow.[50] Brown is quickly aware of the difference between what he came to see and the city he now finds himself in.

If Brown is disillusioned by his initial introduction to Los Angeles, after his transformation into lawyer/spokesperson for the Chicano movement, he gradually becomes aware of the number of inconsistencies and inequalities that make up the larger picture of the city. "The courthouses, the city halls, the county buildings, and federal offices are within a couple of blocks from where I live with cockroaches, winos, pimps, whores, junkies, fags, yoyos with bloody noses and bad breath."[51] The people who run the city from the downtown area are surrounded during their working hours by all the dangerous elements of society but choose to ignore it, as it only costs a nickel to escape from these areas after work to return to the security of their suburban home or sprawling mansion.

On Christmas Eve, 1969, three hundred Chicanos attempt to enter the wealthiest church in Los Angeles where, "from the mansions of Beverly Hills, the Faithful have come in black shawls."[52] The Chicanos are denied entry and the church is blocked by police officers, the supposed sanctuary of a Catholic church becoming a further battlefield to keep the 'cockroaches' out. Brown witnesses another part of the city immersed in a continual struggle as he drives through Watts, "past the bar-b-ques, the liquor stores, the dark alleys." He sees, "the faded buildings are covered with slogans and graffiti the same as in Tooner Flats. It is an older war zone."[53] Largely ignored by the outside world and the majority of Los Angeles before the riots in 1965, it still remains part of war zone, an internal struggle to maintain a community's integrity and fight for equality and recognition.

The media images in *Revolt of the Cockroach People* work in two ways. The importance of presenting the Chicano movement to the outside world is not lost on Brown who takes advantage of the publicity scrum following the group's demonstrations. The world's press, seizing on the novel idea that Chicanos had not "fought inside a temple since the Spanish conquistadores invaded the shrines of Huitzilopochtli in the Valley of Mexico,"[54] turn out in force. Brown stands, "under the glare of hot white lights," and speaks, "to the citizens of Los Angeles and Beirut, to the people in Akron and New Jersey, even to the blokes in England,"[55] ignoring the media, he speaks directly to an international audience. However, the media is a dangerous tool, used by city officials to influence the residents of Los Angeles. Just as *The Black Dahlia's* Ellis Loew used two sporting heroes to improve the public's approval of the elected officials, the thirteen Chicano militants arrested after a school demonstration, "just happened to coincide with the California Primary."[56] Government officials, especially in Los Angeles, are aware of the importance of maintaining the image of order.

While the media has the ability to manipulate the truth and alter the past depending on its bias, personal experience and memory may be changed by the individual. When Brown visits his sister in suburban Los Angeles, he is shocked to find her married to a white, ex-Marine, living in, "a huge breadbox with a white picket fence, a kidney in the back yard for swimming, and a two car garage." Buffalo knows, "it isn't going to work out even before Dave Hurley gets home."[57] He sees his sister, while pursuing her own version

of the American Dream as, "trapped, chained by herself to that blue-eyed fag and his

promise of more make-up and martinis."[58] Brown views her escape from the barrio as a

cop-out, a relinquishment of her history and culture as a Chicano. His nightmare is a

memory from his eighth-grade graduation where Mexican students are not allowed to

march alongside American students. Without the support of his fellow students and family,

and threatened with not graduating, Brown is helpless to make a stand. Years later,

sweating on his bed in Los Angeles and reminded of this incident, he rants against all the

wrong-doing in the world, "the East coast school of books and snow, the Pilgrims and

Baptists...FDR and his quaint limp, Hitler and his ovens, the death of the Incas and the

Aztecs,"[59] represent the frustration and pain of all races and generations, past, present and

future. It is a history that those with power try to eradicate, but it is a constant battle with

the people of the city. The primary senses alone may return an individual to a particular

time or experience. Ruben Martinez, in his essay, 'La Placita' finds a collection of

memories and history invoked through the smell of a family kitchen. "These smells are

leading me back into my own history – an endless spiral of conquests, exiles and

pilgrimages in search of, by turns, the American Dream and Old Mexico...through

revolutions, racism and melding of Norths and Souths – the history erased by the non-

history of growing up in Los Angeles."[60] As much as the histories of other places and cities

are written about, argued and discussed in Los Angeles, the past is suppressed and only the

positive images of its beginnings find their way into the magazines.

In the films *Grand Canyon* (1991) and *Crash* (2004), the viewer sees Los Angeles

as a fragmented and disconnected post-modern spatial area, confused with racial, economic

and cultural tension and desensitized by the constant threat of natural and physical violence.

Lawrence Kasdan's *Grand Canyon* uses the media to connect areas of the city through television. One shot shows an affluent middle-aged white male watching a basketball game in the suburbs, then cuts to a young, black male watching the same game. Kasdan also uses media to highlight the journey of Davis, a producer of violent films, who after being held-up and shot, denounces the very industry that has made him rich as the primary reason for the violence in society. The media have de-sensitised a public grown used to the spectacle of disaster. The desensitization of Los Angeles residents, not only to violence but also to the threat of natural disasters is depicted in Grand Canyon by the behaviour of people during an earthquake. People calmly go through the suggested safety routine during an earthquake, laughing and joking as they do so.

The confusion of racial stereotyping and blurring of difference between right and wrong is explored in Paul Haggis' *Crash*. Opening with two young black men walking in a predominately white area, one of the men complains about being mistreated, judged and viewed with suspicion. He finishes the tirade on injustice by pulling out a gun and car-jacking an expensive SUV. A hard-working Hispanic is accused of being a thief and 'gang-banger' only to be shown later as a devoted father and husband. Both films show corruption, misunderstanding and suspicion are not confined to one area of Los Angeles but exist in every neighbourhood. Fear is seen to be promoted by the popular media, and fuelled by the misconceptions people have of different racial, economic and cultural backgrounds.

Octavia Butler's futuristic L.A. novel, *Parable of the Sower*, set in 2024, "intensifies the contradictions of modern society."[61] It too presents the city as fragmented, and offers a dystopian view of a society in which people are either predator or prey. The earlier visions of Nathanael West's chaotic mob scenes in *The Day of the Locust* is now a reality and the gang mentality and violence depicted in *Menace II Society* has exploded out of South Central to overwhelm residents all over Southern California. Inflation and the demand for goods, and the inadequacy of the police force to combat the continuous

episodes of violent crime, have pushed the privileged into gated communities. The outside

world is hostile, unpredictable and dangerous. Jerry Phillips comments:

> *Parable of the Sower* depicts a harrowing world in which market exchanges and private
> property are the exclusive means of organizing social life. On the one hand, Butler portrays
> certain aspects of late capitalism – its atomistic culture, its elevation of profits over people,
> its volatile race relations, and its ecological destructiveness – as dystopia achieved. On the
> other hand, she intimates that the present has not yet exhausted its barbaric potentialities;
> cannibalism, widespread terrorism, and brutal social repression are the defining
> characteristics of the terrible future that possibly awaits us – a dystopia imagined.[62]

The view is, of course, in direct opposition to the earlier imagined Los Angeles, a city built

around prosperity, growth and healthy living. All of the social factors, the economic

booms, racial tension and ignorance of environmental deterioration have been pushed

beyond their limits to create Butler's world in *Parable of the Sower*. The deception of

security behind a gated community is only a brief respite from the chaotic world, where the

less fortunate will eventually become desperate enough to find a way through the electric

fences and guns.

Simple events and objects taken for granted in the modern world, become both

dangerous and novel in the futuristic community of Butler's Los Angeles. The

domesticated dog is too expensive for the average person to keep; they roam the territory

outside the walls in feral packs. Rain occurs only rarely, people "can remember the rain six

years ago," a momentous occasion with, "water swirling around the back porch."[63] Yet

even rain incites the fear of infection and disease. Guns are an important part of protection,

while, "the police may be able to avenge you, they can't protect you."[64] Children are taught

how to fire a gun at an early age; it is never known when they may have to protect their

family or community from intruders.

The gated communities offer a false and temporary sense of security and are an extension of the contemporary ideas of privatopia. Michael Dear and Steven Flusty comment, "Privatopia has been fuelled by a large dose of privatisation, and promoted by an ideology of 'hostile privatism.'"[65] The large corporations that dominate the city landscape, buying up land and assets, entice people to live and work under their rules and conditions and create an atmosphere of distrust and fear. Defending her family's decision to move to the Olivar gated community in *Parable of the Sower*, Joanne Garfield comments, "It's a better fortress. It won't have people coming over the walls, killing old ladies."[66] Privatopia, "has provoked a culture of nonparticipation,"[67] and Los Angeles by 2024, always a divided city has become separated into small, privatised, fortified out-posts surrounded by the desperate and dying.

The narrator in *Parable of the Sower*, Olamina, attempts to justify hope of a future improvement by creating a new religion, the Earthseed movement. The "destiny of Earthseed is to take root among the stars,"[68] now that hell has moved above ground to take root in Los Angeles. The physical and spiritual journey of Olamina and Earthseed begins when the walled community she lives in is breached. As she escapes "from the neighbourhood it was burning. The houses, the trees, the people: Burning."[69] While people once made the journey West in search of opportunity and a new beginning, Olamina now flees the City of Angels, heading North, leaving behind a burning, destroyed shell of the former powerful city. The future for Olamina and Earthseed is away from Los Angeles, now considered the place where, "most stupid or wicked things began."[70] The imagined city, once revered as a golden land of opportunity, is destroyed by a maniacal mob, corrupted by the prospect of a desolate future.

Olamina's journey North begins as a solitary one, but as she progresses, her Earthseed ideals draw several people closer and a fragile group forms. Offered a block of land by one of the group, Bankole, as a means to start a new life away from Los Angeles, Olamina decides the Earthseed movement will revert back to the agrarian virtues of living off the land. Unable to guarantee security or success, Olamina's and Earthseed's motives still permit a faint sense of hope for the future, and the possibility of a regenerated community, only the future lies far beyond the borders of Los Angeles. The city, in *Parable of the Sower,* has become the apocalyptic reversal of the original California dream. The Pacific Ocean encroaches on the city, eating away at the crumbling earth. The highway becomes obsolete as gasoline prices rise to astronomical prices. Self-preservation and justice are maintained through the accumulation of weapons. Existing borders and boundaries dividing suburbs and towns lose relevance as people are forced to constantly move in order to survive.

The media controlled images that attempt to hide the tension and diversity of Los Angeles behind glorified pictures, also ignore the reality of a city heading towards disaster, too often caught living in an imagined past. The literature and films of Los Angeles react to these images by turning their attention to forgotten parts of the city, choosing to highlight the violence, prejudice and shattered dreams that fill the lives of many of the people who inhabit those communities. The images of Hollywood in Joseph Wambaugh's *The Onion Field,* and of South Central L.A in Walter Mosley's *A Little Yellow Dog,* reveal a city overcome with cultural, racial and economic disparities.

---

[1] Reyner Banham, *Los Angeles – The Architecture of Four Ecologies* (Norwich, UK: Fletcher and Son Ltd., 1971), p. 161.
[2] Edward Soja, *Postmetropolis: Critical Studies of Cities and Regions* (Malden, MA: Blackwell Publishers, 2000), p. 136.
[3] Ibid, p. 136.

[4] Robert MacDougall, 'Red, Brown and Yellow Perils: Images of the American Enemy in the 1940s and 1950s', *Journal of Popular Culture*, 32.4 (1999), 59-73.

[5] Richard Castillo, 'The Los Angeles "Zoot Suit Riots" Revisited: Mexican and Latin American Perspectives', *Mexican Studies* 16.2 (2000), p. 370, in JSTOR <http://links.jstor.org/sici?sici=0742-9797%28200022%2916%3A2%3C367%3ATLA%22SR%E2.0.CO%3B2-6> [accessed 19 February 2007].

[6] *The Los Angeles Riots – Lessons for the Urban Future*, ed. by Mark Baldassare (Boulder, CO: Westview Press, 1994)

[7] James Ellroy, *The Black Dahlia* (Auckland, NZ: Random House Limited, 2005), p. 389.

[8] Scott Bowles, 'Time has yet to fade infamous L.A murder', *USA Today*, 15 September 2006, in *Academic Search Premier* <http://search.ebscohost.com.ezproxy.waikato.ac.nz:2048/login.aspx?direct=true&db=aph&AN=JOE288149 612506&site=ehost-live> [accessed 16 February 2007].

[9] Julia Scheeres, 'Black Dahlia', *Courtroom Television Network 2006* <http://www.crimelibrary.com/notorious_murders/famous/dahlia/2.html> [accessed 20 February 2007].

[10] James Ellroy, *The Black Dahlia*, p. 11.

[11] Douglas Daniels, 'Los Angeles Zoot: Race "Riot," the Pachuco, and Black Music Culture', *The Journal of African American History*, 87 (2002), p. 98, in *JSTOR* <http://links.jstor.org/sici?sici=1548-1867%28200224%2987%3C98%3ALAZR%22T%3E2.0.CO%3B2-I> [accessed 14 December 2006].

[12] James Ellroy, *The Black Dahlia*, p. 393.

[13] Ibid, p. 12.

[14] Ibid, p. 12.

[15] Ibid, p. 28.

[16] Ibid, p. 92.

[17] Ibid, p. 94.

[18] Ibid, p. 277.

[19] Ibid, p. 277.

[20] Carole Allamand, 'A Tooth for a Private Eye: James Ellroy's Detective Fiction', *Journal of Popular Culture*, 39 (2006), p. 357, in *ProQuest* <http://proquest.umi.com/pqdweb?did=1059152661&sid=9&Fmt=3&clientId=8119&RQT=309&VName=P QD> [accessed 14 December 2006].

[21] Joseph Wambaugh, *The Onion Field* (New York, NY: Dell Publishing Co., Inc, 1973), p. 1.

[22] Ibid, p. 1.

[23] Ibid, p. 235.

[24] Ibid, p. 68.

[25] Joan Didion, *Play It As It Lays*, p. 23.

[26] Joseph Wambaugh, *The Onion Field*, p. 72.

[27] Ibid, p. 312.

[28] Ibid, p. 389.

[29] Ibid, p. 231.

[30] Roger Berger, '"The Black Dick": Race, Sexuality, and Discourse in the L.A. Novels of Walter Mosley' in *African American Review*, 31 (1997), p. 285, in *JSTOR* <http://links.jstor.org/sici?sici=1062-4783%28199722%2931%3A2%3C281%3A%22BDRSA%3E2.0.CO%3B2-E> [accessed 10 December 2006].

[31] Helen Lock, 'Invisible Detection: The Case of Walter Mosley', *MELUS*, 26.1 (2001), p. 70 in ProQuest <http://proquest.umi.com/pqdweb?did=86926720&sid=2&Fmt=3&clientId=8119&RQT=309&VName=PQD > [accessed 26 February 2007].

[32] David Fine, *Imagining Los Angeles*, p. 144.

[33] Walter Mosley, *A Little Yellow Dog* (New York, NY: Simon and Schuster Inc., 1996), p. 15.

[34] Ibid, p. 22.

[35] Joseph Wambaugh, *The Onion Field*, p. 149.

[36] Walter Mosley, *A Little Yellow Dog*, p. 77.

[37] Ibid, p. 24.

[38] Ibid, p. 157.

[39] Ibid, p. 148.

[40] Paula Massood, 'Mapping the Hood: The Genealogy of City Space in *Boyz N the Hood* and *Menace II Society*', *Cinema Journal*, 35 (1996), p. 89, in *JSTOR* <http://links.jstor.org/sici?sici=0009-7101%28199624%2935%3A2%3C85%3AMTHTGO%3E2.0.CO%3B2-3> [accessed 10 December 2006].

[41] *Menace II Society*. Dir. Allen and Albert Hughes. New Line Cinema. 1993.

[42] *Menace II Society*. 1993.

[43] Paula Massood, 'Mapping the Hood: The Genealogy of City Space in *Boyz N the Hood* and *Menace II Society*', p. 88.

[44] Raymond Paredes, 'Los Angeles from the Barrio: Oscar Zeta Acosta's *The Revolt of the Cockroach People*' in *Los Angeles in Fiction*, pp. 239-52.

[45] Ruben Martinez, 'La Placita' in *Sex, Death and God in L.A.* (New York, NY: Random House Inc., 1992), pp. 225-58.

[46] Octavio Paz, *The Labyrinth of Solitude* (Auckland, NZ: Penguin Books (NZ) Ltd, 1990), p. 13.

[47] Ibid, p. 13.

[48] Oscar Zeta Acosta, *The Revolt of the Cockroach People* (New York, NY: Vintage Books, 1989), p. 22.

[49] Ibid, p. 23.

[50] Ibid, p. 23.

[51] Ibid, p. 48.

[52] Ibid, p. 11.

[53] Ibid, p. 164.

[54] Ibid, p. 78.

[55] Ibid, p. 79.

[56] James Ellroy, *The Black Dahlia*, p. 54.

[57] Oscar Zeta Acosta, *The Revolt of the Cockroach People*, p. 25.

[58] Ibid, p. 26.

[59] Ibid, p. 31.

[60] Ruben Martinez, 'La Placita', p. 245.

[61] Jerry Phillips, 'The intuition of the future: Utopia and catastrophe in Octavia Butler's *Parable of the Sower*', *Novel*, 35 (2002), p. 300, in *ProQuest* <http://proquest.umi.com/pqdweb?did=403665751&sid=3&Fmt=3&clientld=8119&RQT=309&VName=PQD> [accessed 11 December 2006].

[62] Ibid, p. 302.

[63] Octavia Butler, *Parable of the Sower* (London, UK: The Women's Press, 1995), p. 45.

[64] Ibid, p. 38.

[65] Michael Dear and Steven Flusty, 'Postmodern Urbanism', *Annals of the Association of American Geographers*, 88 (1998), p. 55, in JSTOR <http://links.jstor.org/sici?sici=0004-5608%28199803%2988%3A1%3C50%3APU%3E2.0.CO%3B2-K> [accessed December 10 2006].

[66] Octavia Butler, p. 117.

[67] Michael Dear and Steven Flusty, 'Postmodern Urbanism', p. 55.

[68] Octavia Butler, *Parable of the Sower*, p. 71.

[69] Ibid, p. 141.

[70] Ibid, p. 198.

CONCLUSION

From its earliest beginnings, Los Angeles has been a city built on images of prosperity, fulfilment and opportunity. Yet, too often these images have deflected the gaze from a real presence of a darker side to the city, one of corruption, violence and isolation. In choosing only to promote the redeeming qualities, without acknowledging the problems, city boosters ignored the vast majority of the Los Angeles population. The literature, through its storytelling, exposes the city's past, present and future, through a dark vision of what it had become, a vast and sprawling metropolis that for many people was both a confusing and desperate home. The literature turns from media projected images of Beverly Hills, Malibu Beach and the Hollywood Hills, and demonstrates a reality far-removed from the sun-kissed pictures. Los Angeles is riddled with stories from individuals sharing their fears, disillusionment and broken dreams. Elizabeth Adams comments:

> Los Angeles is a city of stories. The stories get told to define the city for the teller. Here is a place I like, they say. Here is a place I'm scared of, they say. Here is where I was during the riots/uprising, they say. This place was hurt during the earthquake and has not recovered, they say. The city looms and spreads. A single person (or even several) cannot hope to make sense of its vastness.[1]

Los Angeles, so attractive to writers and artists since the birth of film, is a city where people will continue writing and talking about their experiences. The literature takes these experiences and opens up the city in its entirety, sharing the harsh, unpredictable world with the outsider.

As the final frontier for dreamers who travel west, Los Angeles has always appealed as a city of opportunity. But while the road and the sea ignite romantic notions of freedom and adventure, along with the surrounding mountains and desert they serve only to keep Los Angeles penned in against the broad expanse of the Pacific Ocean. In McCoy's *They Shoot Horses Don't They*, images of entrapment abound with the endless circular nature of the dance competition taking place on a pier over the Pacific. The desperation of the competitors and the corruption of the organizers is also reflected in works such as Raymond Chandler's *The Big Sleep*. The hard-boiled detectives expose a darker, more cynical side to the city, where everyone is potentially a fraud. Beneath the shiny surface of a city founded on sunshine and prosperity, corruption reaches all levels of society and there is nowhere left to escape to on the 'mean streets' of Los Angeles.

As the city grows, the collision of technology, rapid progression and population explosion turn Los Angeles into a disconnected city, where the real and imagined merged in a cityscape that demonstrated a conflicting combination of historical replication, original design and movie-set inspiration. The Los Angeles of the literature shows every sign of becoming what Edward Soja describes as an 'Exopolis':

> The Exopolis itself is a simulacrum: an exact copy of a city that has never existed. And it is being copied over and over again all over the place. At its best, the Exopolis is infinitely enchanting, at its worst it transforms our cities and our lives into spin-doctored "scamscapes," places where the real and the imagined, fact and fiction, become spectacularly confused, impossible to tell apart.[2]

The growing confusion of the landscape, where fact and fiction become impossible to separate, is reflected in the literature that reveals its contradictory nature. Nathanael West's Tod Hackett, himself a painter, is driven to comment again and again on a city whose architecture is drawn from every corner of the world — Swiss chalets and Egyptian

pyramids in the heart of California. Characters like Oedipa Maas in *The Crying of Lot 49*

illustrate the overwhelming feeling of powerlessness; as an individual in a society of false

information, she tries to make sense of a comically senseless world.

Alongside the confusion of the 'scamscape' is the shared belief of a growing

number of people in a desperate and empty future. While a Post-war boom generated a

new wave of economic confidence in Los Angeles, the publicised buoyancy was countered

by the oppressive influence of McCarthyism, nuclear depression and loss of identity. Joan

Didion's Maria Wyeth in *Play It As It Lays* dismisses the possibility of Los Angeles as a

city of opportunity and sees only despair and destitution in a city that has come to represent

the angst felt throughout American society. As Lynell George comments:

> There is, despite the abundance of sunshine and wealth, an oppressive sense of doom about
>
> an indifferent and chaotic world. A profound emptiness and a gray despair are both cradled
>
> snugly within this vast lap of luxury. For some, it only becomes increasingly apparent
>
> everyday – there is not the need nor time to fret over a future that may never be.[3]

For some, the chaotic nature of life is simply too much to comprehend, and death is seen as

the only escape. For others, like Allen Ginsberg in 'Howl', the stifling atmosphere

becomes a challenge to break free from the political and cultural shackles restraining

society. Perceptions of tragedy and persecution expressed in 'Howl' reflect not only the

fears of his generation, but of earlier generations of Americans too.

Nothing is ever what it appears to be in the city of Los Angeles. The influence of

Hollywood and the film industry, combined with the non-stop barrage of media images,

continue to blur the line between the real and the imagined. In *Mulholland Drive* (2001)[4], a

beautiful woman stands on an auditorium stage at 2.a.m singing a tragic, haunting song to a

packed house. The audience is moved to tears by the emotion she projects, only to be

shocked as the woman passes out unconscious while the singing continues. The music has been recorded and the woman lip-synching, highlighting again the confusion and impossibility of recognizing the difference between fact and fiction. People dress, eat and talk as though they are living out their lives in front of a movie camera. The city is filled with people going about their mundane existence acting out a very different life, often hiding intense emotions of frustration and anger towards the people who have succeeded. Nathanael West's exploration of Los Angeles in *The Day of the Locust* is filled with these disillusioned performers, all waiting for something to happen to change their lives. In the end, the desperation becomes overwhelming and a chaotic mob forms, destroying the city as a means of filling the emotional void of a meaningless life.

The literature of Los Angeles explores the past, present and the future to highlight a continuing struggle for identity in a city that promised so much but too often fails to meet expectations. That identity was often Mexican, African-American, or native American and the writing of these groups demonstrate individuals who are alienated, confused and desperate. As Paz commented;

> They have lived in the city for many years wearing the same clothes and speaking the same language as the other inhabitants, and they feel ashamed of their origin; yet no-one would mistake them for authentic North Americans.[5]

The desperation is painfully evident in the WASP population too. Alison Lurie's character, Katherine Cattleman, in *The Nowhere City*:

> I don't like the climate. I don't like the sun shining all the time in November, and the grass growing. It's un-natural, it's as if we were all shut up in some horrible big green-house away from the real world and the real seasons. I hate the oranges here as big as grapefruits and the grapefruits as big as, I don't know what, as big as advertisements for grapefruit, without any taste. Everything's advertisements here. Everything has a wrong name, I mean the name for everything, you see, it's always a lie, like an advertisement. For instance, this is Mar Vista, which is supposed to be Spanish for "view of the sea". That's because it has no view of the sea; it's all flat, it has no view of anything. Mar Vista! Spoil-the-view, I call

it; Spoil-the-view, California. I despise it here. You know what I saw the first day I got to Los Angeles, when Paul was driving me back from the airport, the first afternoon I was here? We were driving back from the airport, and we passed a doughnut stand, and sitting on top of it was this huge cement doughnut about twenty feet high, revolting around. I mean revolving. That was the thing I saw, before I saw the stand. From a long, long way off, that big empty hole going around and around in the air, with some name painted on it. Well I thought, that's what this city is! That's what it is, a great big advertisement for nothing.[6]

While the idealised picture of a prosperous city lapping against the Pacific Ocean still excites the imagination of the dreamer and the traveller, the real Los Angeles, as exposed by the literature, can be a much different place, a façade of false, empty promises. The city becomes nothing more than an advertisement for an imagined life, a perfect ocean view or a sun that never stops shining.

---

[1] Elizabeth Adams, 'Making the Sprawl Vivid: Narrative and Queer Los Angeles', *Western Folklore*, 58.2 (1999), p. 179, in *JSTOR* <http://links.jstor.org/sici?sici=0043-373X%28199924%2958%3A2%3C175%3AMTSVNA%3E2.0.CO%3B2-Q> [accessed 14 September 2006]
[2] Edward Soja, Thirdspace – *Journeys to Los Angeles and Other Real-And-Imagined Places* (Cambridge, MA: Blackwell Publishers, 1996), p. 19.
[3] Lynell George, 'City of Specters' in *Sex, Death and God in L.A*, pp. 153-74.
[4] *Mulholland Drive*. Dir. David Lynch. Universal Focus. 2001
[5] Ibid, p. 13.
[6] Alison Lurie, *The Nowhere City* (London, UK: Minerva, 1994), p. 38.

BIBLIOGRAPHY

PRIMARY SOURCES

Acosta, Oscar Zeta, *The Revolt of the Cockroach People* (New York, NY: Vintage Books, 1989)

Anzaldua, Gloria, Borderlands – *La Frontera* (San Francisco, CA: Aunt Lute Books, 1999)

Baldassare, Mark, ed., *The Los Angeles Riots – Lessons for the Urban Future* (Boulder, CO: Westview Press, Inc., 1994)

Banham, Reyner, *Los Angeles – The Architecture of Four Ecologies* (London, UK: Allen Lane The Penguin Press, 1971)

Bowles, Scott, 'Time has yet to fade in famous L.A murder', *USA Today*, 15 September 2006, in *Academic Search Premier* <http://search.ebscohost.com.ezproxy.waikato.ac.nz:2048/login.aspx?direct=true&db=aph &AN=JOE288149612506&site=ehost-live> [accessed 16 February 2007]

Boyz N The Hood, directed by John Singleton, Columbia Pictures, 1991

Bruccoli, Matthew J. and others, ed., *Correspondence of F. Scott Fitzgerald* (New York, NY: Random House Inc., 1980)

Butler, Octavia E., *Parable of the Sower* (London, UK: The Women's Press, 1995)

Capote, Truman, *In Cold Blood* (Harmondsworth, UK: Penguin Books Ltd., 1967)

Chandler, Raymond, *Three Novels* (Harmondsworth, UK: Penguin Books Ltd., 1993)

Clark, William A. V., *The California Cauldron – Immigration and the Fortunes of Local Communities* (New York, NY: The Guilford Press, 1998)

Crash, directed by Paul Haggis, Lions Gate Films, 2004

Davis, Mike, *City of Quartz* (New York, NY: Vintage Books, 1992)

—, *Ecology of Fear – Los Angeles and the Imagination of Disaster* (New York, NY: Vintage Books, 1999)

—, *Prisoners of the American Dream* (London, UK: Verso, 1988)

DeLillo, Don, *White Noise* (New York, NY: Penguin Putnam Inc., 1998)

Dick, Philip K., *Do Androids Dream of Electric Sheep* (London, UK: Grafton Books, 1987)

Didion, Joan, *White Album* (New York, NY: Simon and Schuster, 1979)

—, *Play It As It Lays* (Harmondsworth, UK: Penguin Books Ltd., 1986)

—, *Slouching Towards Bethlehem* (New York, NY: Farrar, Straus and Giroux, 1990)

Dreiser, Theodore, *Sister Carrie* (Philadelphia, PA: University of Pennsylvania Press, 1981)

Ellroy, James, *L.A Confidential* (London, UK: Arrow Books Limited, 1994)

—, *The Black Dahlia* (London, UK: Arrow Books Limited, 2005)

Fitzgerald, F. Scott, *The Last Tycoon* (Harmondsworth, UK: Penguin Books Ltd., 1977)

—, *Tender Is The Night* (New York, NY: Scribner, 1962)

Ginsberg, Allen, *'Howl' and Other Poems* (San Francisco, CA: City Lights Books, 2000)

Grand Canyon, directed by Lawrence Kasdan, 20th Century Fox. 1991

Hammett, Dashiell, *The Maltese Falcon* (New York, NY: Random House Inc., 1989)

Hise, Greg, *Magnetic Los Angeles* (Baltimore, MD: The John Hopkins University Press, 1997)

Huxley, Aldous, *After Many A Summer* (London, UK: Chatto and Windus, 1950)

L.A Confidential, directed by Curtis Hanson, Warner Bros. 1997

Leonard, Elmore, *La Brava* (London, UK: Penguin Books Ltd., 1985)

Lennon, J. Michael, 'A Conversation with Norman Mailer', *New England Review* (Middlebury, NE: 1999), p. 139, in *ProQuest* <http://proquest.umi.com/pqdweb?did=43693347&sid=2&Fmt=3&clientId=8119&RQT=3 09&VName=PQD> [accessed 20 October 2006].

Lurie, Alison, *The Nowhere City* (London, UK: Minerva, 1994)

MacShane, Frank, ed., *The Notebooks of Raymond Chandler and English Summer – A Gothic Romance by Raymond Chandler* (London, UK: Weidenfeld and Nicolson, 1977)

McWilliams, Carey, *The California Revolution* (New York, NY: Grossman Publishers, 1968)

Mailer, Norman, *Advertisements For Myself* (London, UK: Andre Deutsch Ltd, 1961)

—, *The Deer Park* (New York, NY: Howard Fertig Inc., 1980)

Menace II Society, directed by Allen and Albert Hughes, New Line Cinema. 1993

Mizener, Arthur, ed., *F. Scott Fitzgerald – A Collection of Critical Essays* (Englewood Cliffs, NJ: Prentice-Hall Inc., 1963)

Mosley, Walter, *A Little Yellow Dog* (New York, NY: Pocket Books, 1996)

Mulholland Drive, directed by David Lynch., Universal Focus, 2001

Norris, Frank, *McTeague* (New York, NY: The New American Library Inc., 1964)

Paz, Octavio, *The Labyrinth of Solitude* (Harmondsworth, UK: Penguin Books Ltd., 1990)

Pynchon, Thomas, *The Crying of Lot 49* (New York, NY: HarperCollins Publishers Inc., 1999)

Raskin, Jonah, *American Scream* (Berkeley, CA: University of California Press, 2004)

Read, Stephen, Jurgen Rosemann and Job van Eldijk, ed., *Future City*, (Oxon, UK: Spon Press, 2005), pp. 174-93.

Reid, David, ed., *Sex, Death and God in L.A* (New York, NY: Pantheon Books, 1992)

Schulberg, Budd, *The Disenchanted* (London, UK: Allison and Busby, 1983)

—, *What Makes Sammy Run?* (Harmondsworth, UK: Penguin Books Ltd., 1984)

Scott, Allen J. and Edward Soja, ed., *The City – Los Angeles and Urban Theory at the End of the Twentieth Century* (Los Angeles, CA: University of California Press, 1996)

Soja, Edward, *Postmetropolis – Critical Studies of Cities and Regions* (Oxford, UK: Blackwell Publishers, 2000)

—, *Thirdspace – Journeys to Los Angeles and Other Real-And-Imagined Places* (Cambridge, MA: Blackwell Publishers, 1996)

The Maltese Falcon, directed by John Huston, Warner Bros. 1941

Tinker, George E., *Missionary Conquest – The Gospel and Native American Cultural Genocide* (Minneapolis, MN: Orca Press, 1993)

Wambaugh, Joseph, *The Onion Field* (New York, NY: Dell Publishing Company Inc., (1973)

Waugh, Evelyn, *The Loved One* (Bristol, UK: Chapman and Hall, 1958)

West, Nathanael, *Miss Lonelyhearts and A Cool Million* (Harmondsworth, UK: Penguin Books Ltd., 1961)

—, *The Day of the Locust* (Harmondsworth, UK: Penguin Books Ltd., 1963)

SECONDARY SOURCES

Books

Abbott, Megan, *The Street Was Mine – White Masculinity in Hardboiled Fiction and Film Noir* (New York, NY: Palgrave Publishers, 2002)

Adams, Laura, Existential Battles – *The Growth of Norman Mailer* (Athens, Ohio: Ohio University Press, 1976)

Barron, Stephanie and Shari Bernstein and Ilene Susan Fort, ed., *Reading California – Art, Image and Identity, 1900-2000* (Berkeley, CA: University of California Press, 2000), pp. 113-27

Burgh, Patricia, '(De)constructing the Image: Thomas Pynchon's Postmodern Woman' in *Journal of Popular Culture*, 30 (1997), p. 13.

Comerchero, Victor, *Nathanael West – The Ironic Prophet* (Syracuse, NY: Syracuse University Press, 1967)

Cross, K.G.W., *Scott Fitzgerald* (Edinburgh, Scotland: Oliver and Boyd, 1964)

Dugdale, John, *Allusive Parables of Power* (London, UK: MacMillan Press Ltd., 1990)

Fine, David, ed., *Los Angeles in Fiction – A Collection of Essays Revised Edition* (Albuquerque, NM: University of New Mexico Press, 1995)

–, *Imagining Los Angeles – A City in Fiction* (Albuquerque, NM: University of New Mexico Press, 2000)

Hamilton, Cynthia S., *Western and Hardboiled Detective Fiction in America* (Iowa City, IO: University of Iowa Press, 1987)

Higham, Charles and Joel Greenberg, *Hollywood in the Forties* (New York, NY: A.S. Barnes and Company Ltd., 1968)

Horsley, Lee, *The Noir Thriller* (New York, NY: Palgrave, 2001)

Howard, June, *Form and History in America – Literary Naturalism* (Chapel Hill, NC: University of North Carolina Press, 1985)

MacShane, Frank, *The Life of Raymond Chandler* (Toronto, Canada: Clark, Irwin and Company Ltd., 1976)

Madden, David, *James M. Cain* (New York, NY: Twayne Publishers Inc., 1970)

–, ed., *Nathanael West: The Cheaters and the Cheated – A Collection of Critical Essays* (Florida, USA: Everett/Edwards Inc., 1973)

Mendelson, Edward, *Individual and Community: Variations on a Theme in American Fiction* (Durham, NC: Duke University Press, 1975)

Newman, Robert D., *Understanding Thomas Pynchon* (Columbia, SC: University of South Carolina Press, 1986)

Poirer, *Richard, Mailer* (London, UK: Wm. Collins Sons & Co. Ltd., 1972)

Pryce-Jones, David, ed., *Evelyn Waugh and His World* (London, UK: Weidenfeld and Nicolson, 1973)

Shindler, Colin, *Hollywood in Crisis – Cinema and American Society* (London, UK: Routledge Press, 1996)

Tanner, Tony, *Thomas Pynchon* (New York, NY: Methuen and Company Ltd., 1982)

Tytell, John, *Naked Angels* (New York, NY: McGraw-Hill Book Company, 1976)

Watkins, T. H., *The Great Depression – America in the 1930s* (Boston, MA: Little, Brown and Company, 1993)

Articles and Essays

Adams, Elizabeth, 'Making the Sprawl Vivid: Narrative and Queer Los Angeles', *Western Folklore*, 58.2 (1999), p. 179, in *JSTOR* <http://links.jstor.org/sici?sici=0043-373X%28199924%2958%3A2%3C175%3AMTSVNA%3E2.0.CO%3B2-Q> [accessed 14 September 2006]

Allamand, Carole, 'A Tooth for a Private Eye: James Ellroy's Detective Fiction', *Journal of Popular Culture*, 39 (2006), p. 357, in *ProQuest* <http://proquest.umi.com/pqdweb?did=1059152661&sid=9&Fmt=3&clientld=8119&RQT =309&VName=PQD> [accessed 14 December 2006]

Ames, Christopher, 'Shakespeare's Grave; The British Fiction of Hollywood', *Twentieth Century Literature*, 47 (2001), p. 409, in *JSTOR* <http://links.jstor.org/sici?sici=0041-462X%28200123%2947%3A3%3C407%3ASGTBFO%3E2.0.CO%3B2-O> [accessed 7 February 2007]

Baldwin, Kenneth H. and David K. Kirby, ed., *Individual and Community: Variations on a Theme in American Fiction* (Durham, NC: Duke University Press, 1975), pp. 182-222 in *Literature Resource Center* <http://galenet.galegroup.com.ezproxy.waikato.ac.nz:2048/servlet/LitRC?vrsn=3&OP=con tains&locID=waikato&srchtp=athr&ca=1&c=4&ste=16&stab=512&tab=2&tbst=arp&ai= U13033022&n=10&docNum=H1100000676&ST=pynchon%2C+thomas&bConts=278447 > [accessed 29 September 2006]

Berger, Roger, '"The Black Dick": Race, Sexuality, and Discourse in the L.A. Novels of Walter Mosley' in *African American Review*, 31 (1997), p. 285, in *JSTOR* <http://links.jstor.org/sici?sici=1062-4783%28199722%2931%3A2%3C281%3A%22BDRSA%3E2.0.CO%3B2-E> [accessed 10 December 2006]

Callahan, John F., 'F. Scott Fitzgerald's Evolving American Dream: "The Pursuit of Happiness" in Gatsby, Tender Is the Night, and The Last Tycoon', *Twentieth Century Literature*, 42.3 (1996), pp. 374-95, in *JSTOR* <http://links.jstor.org/sici?sici=0041-462X%28199623%2942%3A3%3C374%3AFSFEAD%3E2.0.CO%3B2-J> [accessed 20 July 2006]

Castillo, Richard, 'The Los Angeles "Zoot Suit Riots" Revisited: Mexican and Latin American Perspectives', *Mexican Studies* 16.2 (2000), p. 370, in *JSTOR* <http://links.jstor.org/sici?sici=0742-9797%28200022%2916%3A2%3C367%3ATLA%22SR%E2.0.CO%3B2-6> [accessed 19 February 2007]

Coale, Samuel, 'Didion's Disorder: An American Romancer's Art', *Critique: Studies in Modern Fiction*, XXV.3 (1984), pp. 160-70, in *Literature Resource Center* < http://galenet.galegroup.com.ezproxy.waikato.ac.nz:2048/servlet/LitRC?vrsn=3&OP=conta ins&locID=waikato&srchtp=athr&ca=1&c=7&ste=16&stab=512&tab=2&tbst=arp&ai=U1 3026265&n=10&docNum=H1100002224&ST=didion%2C+joan&bConts=16303> [accessed 1 August 2006]

Daniels, Douglas, 'Los Angeles Zoot: Race "Riot," the Pachuco, and Black Music Culture', *The Journal of African American History*, 87 (2002), p. 98, in *JSTOR* <http://links.jstor.org/sici?sici=1548-1867%28200224%2987%3C98%3ALAZR%22T%3E2.0.CO%3B2-I> [accessed 14 December 2006]

Dear, Michael and Steven Flusty, 'Postmodern Urbanism', *Annals of the Association of American Geographers*, 88 (1998), p. 55, in JSTOR <http://links.jstor.org/sici?sici=0004-5608%28199803%2988%3A1%3C50%3APU%3E2.0.CO%3B2-K> [accessed December 10 2006]

Eagleton, Terry, 'The Illusions of Postmodernism' in *University of Louisville – Fulbright Summer Institute on Contemporary American Literature* (Oxford, UK: Blackwell, 1996)

Geherin, David J., 'Nothingness and Beyond: Joan Didion's Play It As It Lays', *Critique: Studies in Modern Fiction*, XVI.1 (1974), pp. 64-78, in *Literature Resource Center* <http://galenet.galegroup.com.ezproxy.waikato.ac.nz:2048/servlet/LitRC?vrsn=3&OP=con tains&locID=waikato&srchtp=athr&ca=1&c=3&ste=16&stab=512&tab=2&tbst=arp&ai= U13026265&n=10&docNum=H1100002219&ST=didion%2C+joan&bConts=16303> [accessed 1 August 2006]

Glicksberg, Charles I., 'Norman Mailer: The Angry Young Novelist in America', *Wisconsin Studies in Contemporary Literature*, 1.1 (1960), pp. 25-34, in *JSTOR* <http://links.jstor.org/sici?sici=0146-4949%28196024%291%3A1%3C25%3ANMTAYN%3E2.0.CO%3B2-C> [accessed 18 October 2006]

Grenander, M. E., 'The Heritage of Cain: Crime in American Fiction', *Annals of the American Academy of Political and Social Science: 1776-1976*, 423 (1976), pp. 47-66, in *JSTOR* <http://links.jstor.org/sici?sici=0002-7162%28197601%29423%3C47%3ATHOCCI%3E2.0.CO%3B2-5> [accessed 4 July 2006]

Harrison, Barbara, 'Joan Didion: The Courage of Her Afflictions', *The Nation*, 9 (1979), p. 278, in *Literature Resource Center* <http://galenet.galegroup.com.ezproxy.waikato.ac.nz:2048/servlet/LitRC?vrsn=3&OP=con tains&locID=waikato&srchtp=athr&ca=1&c=2&ste=16&stab=512&tab=2&tbst=arp&ai= U13026265&n=10&docNum=H1100002218&ST=didion%2Ctjoan&bConts=16303> [accessed 1 August 2006] (para.3 of 23)

Hausladen, Gary J. and Paul F. Starrs, 'L.A Noir', *Journal of Cultural Geography*, 23.1 (2005), p. 7, in *Academic Search Premier* <http://web.ebscohost.com.ezproxy.waikato.ac.nz:2048/ehost/pdf?vid=3&hid=20&sid=2e1 8eb40-81c5-4a80-b9dc-73c9dde70b45%40sessionmgr2> [accessed 16 February 2007]

Israel, Nico, 'Damage Control: Adorno, Los Angeles, and the Dislocation of Culture', *The Yale Journal of Criticism*, 10.1 (1997), pp. 85-113, in *Muse* <http://muse.jhu.edu.ezproxy.waikato.ac.nz:2048/journals/yale_journal_of_criticism/v010/ 10.1israel.html> [accessed 25 July 2006]

Kroker, Arthur and Marilouise Kroker and David Cook, 'Hypermodernism as America's Postmodernism', *Social Problems*, 37.4 (1990), pp. 443-59, in *JSTOR* <http://links.jstor.org/sici?sici=0037-7791%28199011%2937%3A4%3C443%3APUHAAP%3E2.0.CO%3B2-T> [accessed 28 September 2006]

Levine, Angela, 'The (Jewish) White Negro: Norman Mailer's racial bodies', *MELUS*, 28.2 (2003), in *ProQuest* <http://proquest.umi.com/pqdweb?did=423902091&sid=1&Fmt=3&clientld=8119&RQT= 309&VName=PQD> [accessed 8 November 2006]

Light, James F., 'Nathaniel West and the Ravaging Locust', *American Quarterly*, 12.1 (1960), p. 47, in *JSTOR* <http://links.jstor.org/sici?sici=0003-0678%28196021%2912%3A1%3C44%3ANWATRL%3E2.0.CO%3B2-D> [accessed 19 July 2006]

—, 'Violence, Dreams and Dostoevsky: The Art of Nathanael West', *College English*, 19.5 (1958), pp. 208-13, in *JSTOR* <http//links.jstor.org/sici?sici=0010-0994%28195802%2919%3A5%3C208%3AVDADTA%3E2.0.CO%3B2-G> [accessed 20 July 2006]

Lock, Helen, 'Invisible Detection: The Case of Walter Mosley', *MELUS,* 26.1 (2001), p. 70 in *ProQuest* <http://proquest.umi.com/pqdweb?did=86926720&sid=2&Fmt=3&clientld=8119&RQT=3 09&VName=PQD> [accessed 26 February 2007]

Lord, Geoffrey, 'Mystery and History, Discovery and Recovery' in Thomas Pynchon's *The Crying of Lot 49* and Graham Swift's *Waterland*' in *Neophilologus*, 81 (1997), 145-163

MacDougall, Robert, 'Red, Brown and Yellow Perils: Images of the American Enemy in the 1940s and 1950s' in *Journal of Popular Culture,* 32.4 (1999), 59-73

Massood, Paula, 'Mapping the Hood: The Genealogy of City Space in *Boyz N the Hood* and *Menace II Society*', *Cinema Journal*, 35 (1996), p. 89, in *JSTOR* <http://links.jstor.org/sici?sici=0009-7101%28199624%2935%3A2%3C85%3AMTHTGO%3E2.0.CO%3B2-3> [accessed 10 December 2006].

Phillips, Jerry, 'The intuition of the future: Utopia and catastrophe in Octavia Butler's *Parable of the Sower*', *Novel*, 35 (2002), p. 300, in *ProQuest* <http://proquest.umi.com/pqdweb?did=403665751&sid=3&Fmt=3&clientld=8119&RQT=309&VName=PQD> [accessed 11 December 2006].

Pisk, George M., 'The Graveyard of Dreams: A Study of Nathaniel West's Last Novel, "The Day of the Locust"', *The South Central Bulletin*, 27.4 (1967), p. 65, in JSTOR <http://links.jstor.org/sici?sici=0038-321X%28196724%2927%3A4%3C64%3ATGODAS%3E2.0.CO%3B2-7> [accessed 20 July 2006]

Porter, Joseph C. 'The End of the Trail: The American West of Dashiell Hammett and Raymond Chandler' in *The Western Historical Quarterly*, 6 (1975) p. 411, in *JSTOR* <http://links.jstor.org/sici?sici=0043-3810%28197510%296%3A4%3C411%3ATEOTTT%3E2.0.CO%3B2-1> [accessed 3 July 2006]

Rhodes, Chip, 'Ambivalence on the left: Budd Schulberg's *What Makes Sammy Run*', *Studies in American Fiction*, 30.1 (2002), pp. 65- 84, in *Literature Resource Center* <http://galenet.galegroup.com.ezproxy.waikato.ac.nz:2048/servlet/LitRC?vrsn=3&OP=contains&locID=waikato&srchtp=athr&ca=1&c=4&ste=18&stab=2048&tab=2&tbst=arp&ai=U13696269&n=10&docNum=A96195556&ST=schulberg%2C+budd&bConts=10415> [accessed 4 July 2006]

—, 'The Hollywood Novel: Gender and Lacanian Tragedy in Joan Didion's *Play It As It Lays*', *Style*, 34.1 (2000), in *Academic Premier Search* <http://search.ebscohost.com.ezproxy.waikato.ac.nz:2048/login.aspx?direct=true&db=aph&AN=3667289&site=ehost-live> [accessed 16 February 2007] (para.10 of 39)

Richmond, Lee J., 'A Time to Mourn and a Time to Dance: Horace McCoy's *They Shoot Horses, Don't They?*', *Twentieth Century Literature*, 17 (1971), p. 99, in *JSTOR* <http://links.jstor.org/sici?sici=0041-462X%28197104%2917%3A2%3C91%3AATTMAA%3E2.0.CO%3B2-9> [accessed 12 July 2006]

Romotsky, Jerry and Sally R. Romotsky, 'L.A Human Scale: Street Art of Los Angeles' in *Journal of Popular Culture*, X.3 (1976), 651-59

Rutledge, Gregory E., 'Futurist Fiction & Fantasy', *Callaloo*, 24.1 (2001), pp. 236-52, in *Muse* <http://muse.jhu.edu.ezproxy.waikato.ac.nz:2048/journals/callaloo/v024/24.1rutledge.html> [accessed 25 July 2006]

Schneider, Robert A., 'The Postmodern City from an Early Modern Perspective', *The American Historical Review,* 105.5 (2000), pp. 1668-675, in *JSTOR* <http://links.jstor.org/sici?sici=0002-8762%28200012%29105%3A5%3C1668%3ATPCFAE%3E2.0.CO%3B2-B> [accessed 14 September 2006]

Sides, Josh, 'Straight into Compton: American Dreams, Urban Nightmares, and the Metamorphosis of a Black Suburb', *American Quarterly*, 56.3 (2004), pp. 583-605, in *Muse* <http://muse.jhu.edu.ezproxy.waikato.ac.nz:2048/journals/american_quarterly/v056/56.3sid es.html> [accessed 4 July 2006]

Smethurst, James, 'The Figure of the Vato Loco and the Representation of Ethnicity in the Narratives of Oscar Z. Acosta', *MELUS*, 20.2 (1995), pp. 119-32, in *JSTOR* <http://links.jstor.org/sici?sici=0613-755X%28199522%2920%3A2%3C119%3ATFOTVL%3E2.0.CO%3B2-L> [accessed 11 December 2006]

Tangherlini, Timothy T., 'Los Angeles Intersections (Folklore and the City)', *Western Folklore*, 58.2 (1999), p. 99, in *JSTOR* <http://links.jstor.org/sici?sici=0043-373X%28199924%2958%3A2%3C99%3ALAI%28AT%3E2.0.CO%3B2-1> [accessed 11 December 2006].

Walker, Richard, 'California's Collision of Race and Class', *Representations*, 55 (1996), pp. 163-83, in *JSTOR* <http://links.jstor.org/sici?sici=0734-6018%28199622%290%3A553C163%3ACCORAC%3E2.0.CO%3B2-S> [accessed 12 December 2006]

Wells, Walter, 'Didion's 'Los Angeles Notebook'', *Explicator*, 52.3 (1994), pp. 181-82, in *Literature Resource Center* <http://galenet.galegroup.com.ezproxy.waikato.ac.nz:2048/servlet/LitRC?vrsn=3&OP=con tains&locID=waikato&srchtp=athr&ca=1&c=8&ste=16&stab=512&tab=2&tbst=arp&ai= U13026265&n=10&docNum=H1100002225&ST=didion%2C+joan&bConts=16303> [accessed 1 August 2006]

Widener, Daniel, 'Perhaps the Japanese Are to Be Thanked: Asia, Asian Americans, and the Construction of Black California', *positions: east asia cultures critique*, 11.1 (2003), pp. 135-81, in *Muse* <http://muse.jhu.edu.ezproxy.waikato.ac.nz:2048/journals/positions/v011/11.1widener.html > [accessed 25 July 2006]

Wolff, Cynthia Griffin, 'Didion and the Diver Heroine', *Contemporary Literature*, 24.4 (1983), pp. 480-95, in *JSTOR* <http://links.jstor.org/sici?sici=0010-7484%28198324%2924%3A4%3C480%3A%22IAILD%3E2.0.CO%3B2-U> [accessed 17 September 2006]

Zilberg, Elana, 'Fools Banished from the Kingdom: Remapping Geographies of Gang Violence between the Americas (Los Angeles and San Salvador)', *American Quarterly*, 56.3 (2004), pp. 759-79, in *Muse* <http://muse.jhu.edu.ezproxy.waikato.ac.nz:2048/journals/american_quarterly/v056/56.3zil berg.html> [accessed 25 July 2006]

<u>Websites</u>

Scheeres, Julia, 'Black Dahlia', *Courtroom Television Network 2006*
    <http://www.crimelibrary.com/notorious_murders/famous/dahlia/2.html> [accessed 20
    February 2007]

www.ingramcontent.com/pod-product-compliance
Lightning Source LLC
Chambersburg PA
CBHW072150020426
42334CB00018B/1948